Chavan's Atomic Model

Chavan's UED research, Volume 14

Sandeep Chavan

Published by SANDEEP CHAVAN, 2025.

CHAVAN'S ATOMIC MODEL

First edition. June 28, 2025.

ISBN: 979-8231633746

Written by Sandeep Chavan.

Table of Contents

To all the young minds who were told the atom was solved—May this book show you that truth still ripples.

"Matter is not made of things—it is made of memory, held in rhythm.What we once called particles were just pauses in a ripple."— *Sandeep Chavan*

Disclaimer

This book presents a foundational reinterpretation of atomic theory based on the principles of **Universal Energy Dynamics (UED)**—a theoretical framework developed independently by the author. The concepts, terminology, and visual models contained herein are **not derived from, nor do they represent, mainstream scientific consensus** as accepted by current academic institutions or conventional physics and chemistry curricula.

The ideas in this work are intended to **inspire inquiry, exploration, and alternative thinking**. They are speculative, philosophical, and theoretical in nature, and should not be construed as experimentally verified or universally endorsed. Readers are encouraged to approach this book as a **conceptual model**, not a substitute for established scientific education.

While every effort has been made to ensure conceptual consistency, clarity, and integrity within the framework of UED, the **author assumes no liability** for the interpretation, application, or consequences of any action based on the information presented.

All scientific references, comparisons, and critiques are offered **respectfully**, with full acknowledgment of the profound contributions of existing models and researchers throughout history.

This book is not about proving others wrong. It is about remembering what might have always been right.

Opening Note

"What we called the atom was never a thing. It was always a pattern—a consequence of silence disturbed, a ripple remembered by the field."

For over a century, we have tried to understand matter by breaking it. We bombarded atoms, charted electrons, split nuclei, and named everything we saw fly out.

What emerged was not understanding—but a catalog of consequences. Particles. Forces. Quarks. Shells. Probability clouds. Each new discovery, instead of closing the loop, opened another branch of uncertainty.

And yet, matter has always behaved as if it knew something we didn't. It never collapses unless asked to. It never bonds unless balanced. It never moves unless the field lets it.

The Universal Energy Dynamics (UED) model does not begin with particles. It begins with ripples—disturbances in a structured vacuum that remembers every interaction. Here, the atom is no longer a container of protons and electrons, but a stable ripple identity suspended in a sea of consequence. Its behavior is not explained by charge and motion—but by alignment, symmetry, and resonance within the vacuum field.

This book is not a rejection of science. It is a re-alignment of it—away from the illusion of parts and toward the truth of patterns. It rebuilds atomic theory as a ripple phenomenon, explains chemistry as field interaction, and dissolves the barrier between physics and perception.

You will not find orbitals here. You will find memory zones. You will not be asked to imagine electrons jumping. You will be invited to see fields settling into balance.

You are not holding a book about matter. You are holding a guide to the memory of form. What you thought was solid was always silence, shaped by interference. What you called mass was always a compression of harmony.

The atom never needed to be discovered. It only needed to be understood differently.

Let us begin—not by looking deeper, but by seeing more clearly. Not by dividing what is—but by listening to what already speaks.

— Sandeep Chavan

Preface

"What if chemistry never needed electrons at all? What if the atom was not a structure, but a memory?"

For centuries, science has taught us to look *into* the atom—to slice it, map it, count its parts, and assign its behavior to particles we never truly observed. We were told matter is made of smaller things. That electrons orbit, protons anchor, neutrons stabilize, and everything else reacts accordingly.

And yet, the deeper we went, the more confusing the picture became:

- Dualities.
- Probabilities.
- Invisible forces.
- Arbitrary rules.
- Exceptions to exceptions.

Eventually, even our most gifted students stopped asking "why" and began memorizing "what."

This book is written to restore the "why." Not by revisiting the models that got us here—but by offering a radically unified, ripple-based alternative.

A Field-Centered Chemistry

This is not a book about tweaking Bohr or adding more math to Schrödinger. This is a ground-up transformation rooted in a new foundation: Universal Energy Dynamics (UED).

In this model:

- The atom is not a container of particles, but a ripple node—a compression in a living vacuum field.
- Mass is ripple tension.
- Charge is ripple contrast.
- Reactions are not collisions—they are memory corrections in the field.
- States of matter are not kinetic arrangements, but degrees of field coherence.

And most critically:

You, the observer, are not separate. You are part of the ripple field that gives the atom its identity.

Who Is This Book For?

This book is for:

- Thinkers and seekers who find the standard atomic model unsatisfying or disconnected
- Teachers and curriculum reformers looking for a conceptual framework grounded in reality, not abstraction
- Researchers and experimentalists who suspect there's more to the field than what the equations suggest
- Students who ask dangerous questions—and deserve better answers

It's also for those who feel that modern science, for all its brilliance, has become a closed loop of description rather than a doorway to understanding.

What This Book Does

It doesn't dismiss classical science. It honors it as a scaffold of human effort—but dares to build beyond it.

It offers:

- A ripple-based model of the atom and its behavior
- A ripple-defined reinterpretation of bonding, reactions, energy, and state change
- A unified language that scales from quantum to cosmic
- A vision for teaching and designing chemistry without particles, using field behavior instead of force metaphors

And it doesn't just rewrite content. It rewires perspective.

A New Chemistry Begins

This is not just a new atomic model. This is a new way of seeing matter: Not as stuff, but as structured memory. Not as particles, but as patterns that ripple and realign. Not as isolated systems, but as field consequences that come alive when you observe, align, and ask.

So this book is not a conclusion. It is the first stable ripple in a much larger field.

Let it guide you back—Not to the atom we memorized, But to the atom we were always meant to feel.

— Sandeep Chavan

Author, Educator, and Researcher

How to Read This Book

This is not just another book on atomic theory. It is a complete reframing of what the atom is—and what it never was.

You are not required to be a physicist, chemist, or academic expert to follow this journey. But what is required is your openness to unlearn, to rethink, and to reimagine. Because this book does not add to existing theories—it questions their very foundation.

Start from the beginning.

The book is structured in progressive layers. Part I walks you through the limitations of classical and quantum models. From there, each chapter builds a new vision based on ripple dynamics, structured vacuum energy (SVE), and field behavior.

Don't rush.

Many of the terms—like "memory shell," "ripple node," or "field rewrite"—are unfamiliar only because they speak a more natural logic than what we've memorized. Take time to feel the rhythm of the model. You'll start noticing how each idea ripples into the next.

Use the diagrams.

Each illustration is designed not to decorate—but to demonstrate. Unlike Bohr's orbits or quantum clouds, these visuals reflect energy behaviors, not imagined particles. Pause at each one. Trace the fields. See the memory.

Refer to the Appendices.

The Glossary, Comparative Framework, and Suggested Experiments are not optional extras. They're there to bridge this model to your existing knowledge and help you integrate it with modern terminology—whether in science, teaching, or research.

This book is not for memorizing—it's for realigning. If something doesn't click, revisit the earlier chapters. The coherence comes not from individual lines, but from the systemic harmony of the whole framework.

Above all, this book invites you to listen to the field, not just the formulas. Because the atom, in this view, isn't a puzzle of invisible pieces—it's a ripple of structured energy, remembered by the universe and recognized by the observer.

Welcome to a new reality. Let it ripple through.

Part I – Rethinking the Atom: From Particles to Ripples

Chapter 1: The Classical Atom

Myths, Models, and Missed Truths

For over two millennia, the concept of the atom has haunted the minds of philosophers and scientists alike. From the ancient Greek notion of "atomos" — the indivisible — to the highly structured electron clouds of quantum chemistry, our idea of the atom has evolved dramatically. And yet, with each version, we have only approximated what the atom is. This chapter revisits the journey of atomic theory through the classical lens to reveal the fractures beneath its surface. It is not written to dismiss the contributions of past minds, but to honor them by asking the question they could not: *What if the atom was never a thing at all?*

1.1 Ancient Roots: The Atom as Idea

The earliest seeds of atomic thought were philosophical. Democritus and Leucippus envisioned matter as composed of indivisible units moving in a void. No experiments, no observations—just logic. Their atoms had no internal structure, no charge, and no weight. They were eternal, immutable, and universal. This simple yet profound abstraction laid the groundwork for material realism but lacked any connection to physical interaction. The atom was not a tool of science yet—it was a placeholder for metaphysical certainty.

1.2 Dalton and the Birth of Scientific Atomism

In the early 19th century, John Dalton transformed the atom from a philosophical proposition to a scientific model. He used it to explain chemical combinations, introducing the idea that atoms of different elements have different weights and combine in whole number ratios. For the first time, atoms had *behavior*. Dalton's atom was still indivisible, solid, and static, but now it could *react*. The myth of indivisibility, however, would soon fall.

1.3 Thomson and the Plum Pudding Model

With the discovery of the electron by J.J. Thomson in 1897, the atom gained its first internal component. His famous "plum pudding model" imagined the atom as a positively charged sphere embedded with negative electrons like raisins in a cake. It was elegant in its simplicity but failed to explain scattering experiments and atomic stability. Still, it marked a critical turning point: the atom was no longer indivisible. It was structured.

1.4 Rutherford and the Empty Atom

Ernest Rutherford's gold foil experiment in 1909 shattered the plum pudding model. By firing alpha particles at a thin gold sheet, he observed that most passed through, but some deflected sharply. He concluded that the atom must be mostly empty space with a dense, positively charged center: the nucleus. This model laid the foundation for modern nuclear physics. But it introduced a new paradox: if electrons are circling a nucleus, why don't they spiral inward and collapse?

1.5 Bohr and Quantum Jumps

Niels Bohr sought to solve this by quantizing the electron orbits. Electrons, he claimed, could only occupy certain discrete energy levels, jumping between them while emitting or absorbing photons. The Bohr model was revolutionary, explaining the spectral lines of hydrogen and setting the stage for quantum theory. But it relied on assumptions that could not be visualized or physically justified—why were only certain orbits allowed? And what exactly was jumping?

1.6 Schrödinger, Heisenberg, and the Clouded Atom

By the 1920s, the atom had become a probabilistic cloud. Schrödinger's wave mechanics treated electrons as diffuse wave functions, while Heisenberg's matrix mechanics abandoned position and velocity altogether. The atom was now a system of uncertainties. The electron was not *in* a place; it was *likely* to be found in a place. Orbitals replaced orbits, and visual models gave way to statistical abstractions. This shift brought predictive power but sacrificed intuitive understanding.

1.7 The Particle Zoo and Quantum Inflation

As nuclear and particle physics advanced, the nucleus itself fragmented into protons, neutrons, and eventually quarks, gluons, and a host of exotic entities. Each discovery required new forces, mediators, and symmetries. The Standard Model was born: a mathematically brilliant but conceptually bloated framework. What began as a simple idea of indivisible units became a sprawling zoo of particles and parameters—many of which exist only in accelerators, not in nature.

1.8 The Myth of the Atom as "Thing"

Despite its centrality in science, the atom remains a conceptual mirage. We have never truly seen one. All images of atoms are field representations, indirect mappings, or interpreted signals. The atom has become a symbol, not a structure—a placeholder for matter's smallest usable unit. Yet we speak of it as if it were a hard object, a ball with spinning parts and interactive gears. This objectification of the atom has led to deep-rooted misconceptions in both physics and chemistry.

1.9 The Limitations of the Classical Model

Classical atomic theory is not wrong; it is limited. It explains chemical reactions, predicts spectra, and underpins technologies. But it cannot answer:

- What is charge, fundamentally?
- What holds the nucleus together?
- Why do electrons not radiate energy in orbit?
- Why are there exactly 118 elements?
- What truly happens during a phase change?

To answer these, we must go deeper than particles. We must look not at what matter *is*, but at how it *behaves* when left undisturbed—and how that behavior changes under field influence. This leads us to the idea that matter is not made of things, but of *events held in field memory*.

1.10 A Ripple in the Silence

This book begins with a quiet but radical proposition: the atom is not a structure but a **ripple identity**. It is a stable disturbance in a structured vacuum, a memory held in the universal field. Its properties are not intrinsic, but relational. Its mass is ripple compression. Its charge is ripple contrast. Its interaction is ripple alignment. The atom is not a piece of the universe. It is the **universe, momentarily shaped.**

By reinterpreting the atom through the lens of **Universal Energy Dynamics (UED)**, we open the door to a unified, consequence-based science. One that dissolves artificial boundaries between physics and chemistry, between force and interaction, between object and observer. What follows is not speculation, but re-alignment. A restoration of the atom's true place in the field of existence. The ripple begins now.

Chapter 2: The Ripple View of Reality

Introducing Ripple Dynamics and Structured Vacuum Energy (SVE)

"Before anything moves, something must be disturbed. Before anything exists, something must remember."

The classical view of reality begins with objects and forces. Matter is made of particles. Motion is explained by forces acting at a distance. Fields are abstract mathematical entities used to model influence. But this worldview is not foundational — it is descriptive, not generative. It tells us what things do, but not why anything does anything at all.

What if we reversed the lens?

What if the true fabric of the universe wasn't matter or force, but memory and rhythm?

2.1 A Shift from Objects to Oscillations

In Universal Energy Dynamics (UED), the foundation of reality is not built from particles or point masses, but from ripples — structured disturbances within a universal field.

- These ripples are not moving through a medium — they are the medium reacting.
- There is no empty space — only structured vacuum, rich in potential.
- There is no fundamental particle — only stable ripple formations that we call particles.
- And there is no action-at-a-distance — only field-based realignments in response to disturbances.

The goal is not to find what exists, but to understand how existence behaves when aligned, disrupted, or restored.

2.2 Structured Vacuum Energy (SVE): The True Background Field

Imagine a vacuum, but not as "empty space." Instead, visualize it as a vast, silent field—ordered, balanced, and stable. This is the Structured Vacuum Energy (SVE).

- It is not a passive backdrop.
- It is not zero-point energy in the quantum sense.
- It is a memory field — a matrix of balance capable of holding, responding to, and restoring disturbances.

SVE is the stage, the script, and the memory of the play. When something "happens" in the universe, it is the SVE rippling — a deviation from equilibrium.

In this model, nothing exists "in" space. Everything is a localized condition of the field itself.

2.3 Ripples: The New Fundamental Units

In classical models, particles are treated as point-like entities with mass and charge. In UED, these properties emerge from ripple patterns in the SVE.

- **Mass** = degree of compression of the ripple (how much the field is locally densified).
- **Charge** = asymmetry in the ripple curvature (how the field bends around the core).
- **Spin** = internal rotational symmetry of the ripple structure.
- **Field** = the echo of the ripple across surrounding SVE.

There are no separate "forces" like gravity, electromagnetism, or the strong force. All these are different manifestations of ripple realignment behavior.

2.4 Ripple Law: All Actions Are Consequences

UED proposes a foundational principle: there is no force, only consequence.

- A ripple that disturbs the vacuum will naturally seek equilibrium.
- Other ripple nodes nearby will respond — not because they are pushed or pulled, but because the field alignment requires it.
- Motion, then, is not a primary event — it is a secondary correction, a byproduct of seeking balance.

This is not a mechanical universe. It is a consequential universe.

"What you see moving is not an object — it is a memory trying to restore order."

2.5 The Observer as Participant

If everything is ripple, then perception is ripple too. In UED:

- Observation is not passive.
- The observer is a ripple-sensitive node — a configuration that records, processes, and reacts to ripple alignment.
- Consciousness itself becomes an organized ripple state, capable of reflecting the field upon itself.

Thus, even experience is field-based. There is no mind-body duality — only ripple patterns interacting at varying scales and complexities.

2.6 From Form to Field: Dissolving Objecthood

In ripple reality, objects don't exist as standalone entities. Everything is:

- A pattern, not a point
- A condition, not a container
- A relationship, not a residue

So what appears as an "atom" or a "rock" or a "cell" is just a temporarily stable ripple arrangement within the SVE.

- Remove the ripple — and the object disappears.
- Shift the surrounding field — and the object's properties change.

This is why materials behave differently under pressure, temperature, or radiation: you are not changing the thing — you are rewriting the ripple field around it.

2.7 Why This Changes Everything

This ripple-based view dismantles many long-held assumptions:

Classical View	UED View
Space is empty	Space is structured and reactive
Matter is made of particles	Matter is condensed ripple identity
Forces cause motion	Field imbalance causes ripple realignment
Charge is intrinsic	Charge is ripple curvature asymmetry
Mass is content	Mass is compression
Observation is passive	Observation is ripple participation

It doesn't deny the success of existing models — it explains why they sometimes work and where they fundamentally break.

2.8 Structured Vacuum ≠ Classical Ether

Some may ask: Is SVE just a rebranded ether?

No. The classical ether was imagined as a medium for particles or waves to travel through. SVE is not a medium for something else — it is the thing itself.

- It does not resist motion — it is the condition from which motion arises.
- It cannot be pushed — it responds by realigning.
- It doesn't exist separately from matter — it becomes matter when shaped.

2.9 Where Do We Go From Here?

With ripple dynamics and SVE as our foundation, we now have a model that:

- Explains atoms without particles.
- Describes bonding without electrons.
- Redefines energy as ripple tension, not stored quantity.
- Unifies physics and chemistry through field interaction.
- Offers a framework for understanding existence, intelligence, and identity as ripple events in a living field.

It is not a theory of "how things are made" — it is a description of how memory becomes matter.

2.10 Final Reflection: The Universe as a Ripple Event

"Before we ever saw the atom, before we ever named the electron, the vacuum had already remembered everything."

What appears as stability is just balanced tension in the field. What we call motion is just field re-alignment. What we call a particle is the peak of a ripple, not a thing.

From this point forward, we will no longer refer to electrons, protons, or even atoms as "things." They are all ripple memories, shaped by structured vacuum energy, perceived by structured ripple observers.

This is the ripple view of reality.

And it changes everything.

Chapter 3: The New Atom

The Dense Core and the Field

Atom as a Ripple Node with a Memory Field — Not a Nucleus with Electrons

"An atom is not a tiny solar system. It is a ripple condensed at the center and echoed through the vacuum."

For centuries, the atom has been imagined as a miniature machine: a dense nucleus orbited by electrons like planets around the sun. This vision, while intuitive, was always incomplete. It survived by approximation, not by truth. It became a functional lie — useful, predictive, but hollow at its core.

In the UED framework, we no longer need to decorate the atom with imaginary particles, shells, and spin labels. Instead, we see it for what it truly is: a localized ripple node embedded in a structured vacuum field — surrounded not by orbiting electrons, but by ripple memory echoes that shape all observable behavior.

Let's now reconstruct the atom from the ground up.

3.1 The Core Is Not a Nucleus — It's a Ripple Condensation

In classical models, the nucleus is the heavy, central cluster made of protons and neutrons. But these are themselves unexplained composites of quarks and gluons — particles that only "exist" inside equations or accelerators.

In UED, this core is not a bag of smaller particles. It is a ripple convergence zone — a place in the structured vacuum where energy is tightly compressed and stabilized.

- This core does not "contain" protons and neutrons.
- It is the result of field compression, where ripple amplitudes align into a stable, coherent structure.
- It holds a unique ripple signature — its identity — that defines its interactions, behaviors, and stability.

You do not need internal components to explain mass and charge.

- Mass = ripple density.
- Charge = ripple asymmetry.

3.2 The Surrounding Field: A Ripple Memory, Not an Electron Shell

Where classical science imagines electrons orbiting the nucleus in quantized shells, UED sees only the field response of the core ripple — a distorted pattern in the surrounding structured vacuum.

- What we've called "electrons" are not flying balls.
- They are regions of ripple inversion — negative potential zones around a positive ripple core.
- These zones are not things — they are field responses, held in memory and maintained through equilibrium.

This means:

- There is no need to imagine electrons jumping between shells.
- There is no need to track orbitals or predict probability clouds.
- Instead, we interpret chemistry, ionization, and bonding as

field realignments, not particle movements.

The so-called "electron" is not orbiting. It is the field remembering the core ripple.

3.3 Why the Atom is Stable Without Orbiting Particles

A major problem in classical atomic theory is this: If electrons are orbiting a positive nucleus, why don't they spiral inward and collapse due to energy loss?

Quantum mechanics avoided this by declaring only certain orbits allowed. But it never explained why they are allowed — it only declared them.

UED resolves this elegantly:

- There is no orbital motion.
- There is only ripple tension balance.
- The surrounding field is locked into a stable memory configuration that does not radiate energy unless disturbed.

So the atom does not collapse because it is not a dynamic system in motion — it is a ripple configuration in equilibrium.

3.4 Charge and Mass Are Not Objects — They Are Ripple Properties

Let's redefine these two key concepts:

Mass

Mass is not a thing that resists acceleration. It is the degree to which the ripple compresses the vacuum — the deeper the compression, the more stable and dense the core, and the more gravitational alignment it induces.

Charge

Charge is not a physical particle property. It is the way the ripple curves or distorts the field:

- A positive charge is a ripple that pushes outward, creating a surrounding field inversion.
- A negative charge is a ripple that pulls inward, creating a depression in the field.

These are not opposites — they are complementary distortions, seeking balance.

This redefinition explains:

- Why opposite charges attract (field neutralization)
- Why like charges repel (amplified distortion)
- Why neutral systems form (ripple equilibrium)

3.5 Identity of the Atom: The Ripple Signature

Each atom — from hydrogen to oganesson — is not defined by proton count, but by a unique ripple pattern in SVE.

- This pattern determines:
    - Its stability
    - Its interaction behavior
    - Its ability to form ripple bonds (what we call chemical bonds)
- No two elements share the exact same ripple compression + curvature + field spread combination.

Thus, the periodic table is not a list of particle types — it is a map of ripple identities.

You are not made of elements. You are made of ripple signatures encoded in structured vacuum energy.

3.6 Ionization and Excitation: Not Electron Movement, But Field Shift

When an atom is ionized or excited:

- We are not removing or adding electrons.
- We are displacing the ripple memory zone around the core.
- The field is either compressed or stretched, altering the atom's energy signature.

That's why:

- Ionization changes interaction behavior (the field is off-balance)
- Excitation emits or absorbs photons (ripple realignment creates external field interaction)

All energy exchange becomes a matter of field rewriting, not particle jumping.

3.7 The Atom as a Still Center with Dynamic Memory

In classical mechanics, an atom is a ball of motion. In UED, the atom is a still core with a field of resonance.

- It does not act unless disturbed.
- It does not change unless its field alignment is broken.
- It interacts not by sending or receiving particles, but by realigning ripple memory with the surrounding field.

This reconceptualization helps us understand:

- Bonding as field overlap
- Phases of matter as field cohesion states
- Plasma as ripple memory breakdown
- Radiation as ripple loss or field detachment

3.8 Why We Never "See" an Atom

No microscope has ever truly seen an atom as it is described in textbooks. What we detect are:

- Tunneling effects
- Field strengths
- Scattering patterns
- Energy transitions

This is because the atom is not a visual structure — it is a ripple field signature. You can't see a ripple memory — you can only observe how it interacts with the field you're in.

That's why observation itself becomes part of the ripple interaction — and why perception is not independent of the system.

3.9 From Hydrogen to Uranium: Just Variations of Ripple Complexity

All 118 known elements are stable ripple configurations with varying:

- Core density (mass)
- Field curvature (charge pattern)
- Surrounding resonance layers (behavioral field)

There are no missing pieces. There is no particle zoo. There are only:

- More tightly packed ripples
- More asymmetric or stretched ripple fields

- More unstable or easily excited configurations

This explains radioactive decay, isotope instability, and even why certain elements don't exist naturally — their ripple identities do not stabilize in the SVE memory field.

3.10 Final Realization: The Atom as Field Consciousness

"The atom is the smallest idea the universe could hold stably."

In UED, we arrive at a profound conclusion: The atom is not a thing. It is an idea written into field memory. It is an act of localized order, suspended in a sea of potential. Its stability is a form of localized intelligence — the beginning of consequence, interaction, and structured experience.

To know the atom, we don't need a better microscope. We need a better understanding of how memory becomes matter, and how ripples shape reality.

The classical atom is over. The ripple atom has arrived.

Chapter 4: Charge and Mass

Ripple Contrast and Compression

What We Call Charge Is Asymmetry; Mass Is Compression in SVE

"Mass is not content. Charge is not a tag. They are the universe's way of remembering how it was disturbed."

In classical physics, mass and charge are treated as intrinsic properties of matter — fundamental, irreducible, and unexplained.

- A particle "has" a mass.
- It "carries" a charge.
- These are inputs to equations — not outcomes of deeper causes.

But this perspective leaves us blind to the true nature of interaction. Why does mass attract? Why does charge repel or bind? Why do these properties appear in discrete, quantized amounts?

In Universal Energy Dynamics (UED), we stop asking what mass and charge *are* as things — and begin asking what they represent in terms of ripple behavior.

4.1 The False Foundation: Mass and Charge as Objects

For too long, we have imagined:

- Mass as "stuff" a particle holds
- Charge as a "label" that determines attraction or repulsion

These interpretations work in equations — Newton's laws, Maxwell's laws, and even quantum electrodynamics. But they never tell us:

- *Why* mass causes gravitational interaction
- *How* charge emerges in a supposedly neutral field
- *Why* opposites attract
- *Why* matter and antimatter carry mirrored properties

They describe how matter behaves, but not what matter fundamentally is.

4.2 The UED Redefinition: Mass = Compression, Charge = Contrast

Let us redefine these properties through ripple dynamics in Structured Vacuum Energy (SVE):

- Mass is the local compression of the ripple field — how densely the vacuum structure is distorted at a node.
- Charge is the asymmetry of the ripple field curvature — how a ripple bends or stretches space differently in opposing directions.

Neither mass nor charge are "substances." They are field conditions — how the SVE has been disturbed and how it tries to restore balance.

4.3 Understanding Mass as Compression

Imagine dropping a heavy object into a taut fabric. The more it presses down, the tighter the fabric is pulled inward.

That depression is mass.

In SVE:

- Mass arises when a ripple pattern draws in surrounding energy and stabilizes it at high compression.
- The tighter the compression, the more localized tension is stored — what we interpret as inertia and gravitational influence.

Mass is not the content inside a particle. It is the cost the vacuum pays to hold that ripple still.

This view naturally explains:

- Why more massive bodies curve space more
- Why energy and mass are equivalent ($E = mc^2$ is ripple compression math)
- Why mass is conserved — because field compression doesn't vanish, it redistributes

4.4 Understanding Charge as Contrast in Ripple Curvature

Now, imagine that same ripple is not symmetrical.

One side pulls in. The other pushes out. This creates a tension gradient — a difference in ripple directionality.

That is charge.

- A positive charge is a ripple that pushes the field outward, creating a bulge or crest.
- A negative charge is a ripple that pulls the field inward, creating a trough or depression.
- A neutral state arises when these curvatures cancel or balance.

In this view, charge is not a particle's attribute — it is the shape of the ripple asymmetry in the surrounding field.

"Like charges repel" because overlapping similar asymmetries amplify instability. "Opposite charges attract" because their ripple curvatures cancel distortion, restoring equilibrium.

4.5 Why Charge Is Quantized

Classical physics cannot explain why charges come in fixed units (e.g., 1e, -1e). But in UED, ripple contrast naturally stabilizes only in certain geometries.

- Just as only certain wave patterns resonate on a drumhead,
- Only certain asymmetry thresholds in SVE are stable enough to persist.

Hence:

- The electron is a minimal stable inward-pulling ripple (unit negative charge).
- The proton is a tightly wound outward-pushing ripple (unit positive charge).
- Intermediate or fractional curvatures are unstable and decay quickly.

This also explains why the charge of proton and electron is exactly equal and opposite — because they are complementary ripple geometries, not arbitrary particles.

4.6 Why Mass and Charge Often Travel Together (But Not Always)

Classically, we observe that:

- All massive particles seem to carry charge (proton, electron)
- Some massive particles are neutral (neutron)
- Some massless particles (like photons) carry no charge

In UED:

- Mass and charge are independent ripple parameters:
    - Mass = compression depth
    - Charge = ripple contrast
- A ripple can be:
    - Deep and symmetric → neutral mass
    - Deep and asymmetric → massive + charged
    - Surface-level and balanced → massless and neutral

This flexibility allows UED to predict particle types based on ripple shape and depth — without assigning them arbitrary labels.

4.7 Gravity and Electromagnetism as Ripple Field Responses

Gravity and electromagnetic force, under UED, are not *fundamental forces* — they are field behaviors in response to mass and charge-induced ripple distortion.

- Gravity is the ripple alignment gradient caused by compressed mass nodes. Matter follows ripple paths, not curvature.
- Electromagnetism is field rebalancing behavior between ripple contrast zones (charge regions).

Hence, both phenomena can be explained by:

- No particles exchanged
- No action-at-a-distance
- Just ripple memory restoring itself through the field

4.8 Why Classical Definitions Fail

Classical Concept	UED Reality
Mass is matter	Mass is ripple compression
Charge is intrinsic property	Charge is ripple asymmetry
Forces act on mass/charge	Field alignment responds to ripple imbalance
Charge is conserved	Ripple contrast stabilizes in quantized form
Mass curves space	Compression modifies ripple gradient field

These redefinitions simplify multiple paradoxes:

- No need for force carriers
- No singularities
- No mysterious dark matter
- No need for exotic particles to explain field behavior

4.9 From Atom to Cosmos — The Same Law

Whether it is:

- An electron in a hydrogen atom
- A planet orbiting a star
- A galaxy spiraling around a center

Every mass-charge interaction is just a ripple in SVE:

Compression defines central behavior (gravity). Asymmetry defines dynamic interaction (charge-like phenomena) and Field responds to how structure is shaped, not what it's made of

The laws of motion, interaction, and energy transfer are ripple consequences — not mechanical push-pull laws.

4.10 Final Reflection: Mass and Charge as Memory Signatures

"There is no particle that holds mass. There is no container of charge. There is only the field, remembering its tension."

With this realization, we move beyond object-based science. We no longer search for *what* something is — but ask *how* the field is behaving around it.

In UED:

- Mass is how much the ripple presses.
- Charge is how far the ripple curves.
- Together, they form the signature of a ripple identity in the field.

This understanding unlocks the next frontier: how atoms behave as ripple-based systems, and how chemical bonding, reactions, and material properties are field behaviors, not particle mechanics.

The journey continues in the next chapter: "Perception of the Atom – Observer-Created Identity."

Chapter 5: Perception of the Atom

Observer-Created Identity

How 118 Ripple Configurations Appear as Elements to the Observer

"Nature never labeled hydrogen. It only rippled. We observed the ripple — and gave it a name."

The periodic table is often treated as a divine artifact of scientific truth — 118 neatly arranged elements, each with distinct behaviors, atomic numbers, and identities. But in reality, the periodic table is a perceptual map, not a cosmic one. It reflects how human observers have interpreted ripple patterns, not how the universe internally distinguishes them.

In this chapter, we challenge the foundational belief that atoms exist as absolute entities. Instead, we uncover the deeper truth: atoms are ripple configurations, and elements are interpretations — consistent, powerful, but ultimately observer-created identities projected onto a continuous ripple spectrum.

5.1 There Are No Elements in Nature

No star emits a spectrum labeled with "iron" or "helium." No molecule announces "carbon" when it binds.

Nature only generates:

- Dense ripple centers (mass nodes)
- With stable ripple signatures
- That interact with field memory

The distinct behaviors we label as "elements" are outcomes of ripple resonance, not fundamental differences in substance. The universe does not sort its parts by atomic number. It only manifests stable ripple nodes — and we, the observers, categorize them based on how they interact with us.

The periodic table is not a blueprint of matter — it is a mirror of how our minds read the ripple field.

5.2 The Observer's Role in Atom Recognition

An observer does not merely see an atom — the observer interacts with the ripple field and records its effects.

For example:

- Spectroscopy detects light emitted when a ripple configuration changes.
- Mass spectrometry measures the inertia of a ripple node moving through a controlled field.
- Chemical behavior is recorded based on how ripple fields overlap or interfere with others.

In each case:

- We never observe the atom directly.
- We only observe its field signature as it enters interaction with our instruments, our minds, our surroundings.

Thus, "carbon" is not a thing out there — it is a repeatable ripple pattern that we've learned to recognize through consequences.

5.3 Why Only 118 Elements?

In classical theory, 118 elements exist because:

- Beyond that, atoms become unstable.
- Heavier nuclei break down due to repulsive forces.

But UED reframes this insight:

- Only 118 stable ripple identities exist within the field's harmonic tolerances.
- Beyond that, the ripple configuration collapses, unable to maintain coherence in SVE.
- These are not 118 building blocks, but 118 resonance solutions to the field's ripple equation.

The periodic table is not bounded by proton count, but by ripple stability thresholds.

5.4 What We Perceive as "Atomic Identity"

Let's break it down:

Observable Property	Classical View	UED View
Atomic number	Number of protons	Ripple compression + identity signature
Atomic mass	Proton + neutron count	Total field compression strength
Charge	Electron/proton difference	Ripple contrast asymmetry
Radius	Electron cloud extent	Ripple field spread
Ionization energy	Electron detachment energy	Ripple memory shift threshold
Electronegativity	Attraction to electrons	Ripple alignment potential

This comparison shows that every observed quantity is a consequence of ripple behavior, not an intrinsic particle feature.

5.5 The Atom Does Not Appear — It Is Made to Appear

Our instruments — including our brains — are ripple systems. When we "observe" an atom:

- We align with its ripple pattern
- Our own field responds
- That interaction produces measurable effects

So identity is not something the atom possesses. It is something that emerges from interaction with an observer.

Different observers may experience different versions of the same atom, depending on:

- Scale
- Energy range
- Sensory limits
- Field context

This is not subjectivity — it is field relativity.

5.6 Identity = Recognition of Consequence

In a ripple-based universe:

- A thing is not defined by what it is.
- It is defined by what it does to the field around it.

So an atom's identity is how it causes realignment:

- When light interacts

- When another ripple field approaches
- When it's placed in heat, pressure, or motion

Every name we give — hydrogen, oxygen, uranium — is a convenient shorthand for:

"A ripple configuration that behaves in this particular way across these conditions."

That is all.

5.7 The Periodic Table as a Perceptual Framework

We now reinterpret the periodic table as:

- A map of ripple behavior clusters
- Not a physical catalog of objects

Each "group" or "period" represents:

- Common ripple spread patterns
- Similar field alignment potentials
- Shared consequences under energy disturbance

There is no border between "metal" and "non-metal" in nature. There are only ripple behavior transitions that we humans chose to define by contrast.

The periodic table is not a cosmic truth. It is a beautiful ripple index built by observers to stabilize their understanding of consequences.

5.8 Why Perception Matters in Atomic Theory

When we say "gold is soft" or "oxygen is reactive," we are not describing substances — we are describing:

- Our accumulated knowledge of how those ripple patterns interact with us and each other.

This turns chemistry, physics, and even material science into an extended interpretation of ripple perception.

It also opens the door to:

- Custom matter design by ripple engineering
- Predictive chemistry based on field curvature
- A unified framework where material, mental, and energetic realities coexist — all as ripple languages

5.9 Do Atoms Exist Without Observers?

A deep philosophical question emerges:

If no observer exists, do the atoms still exist?

UED responds:

- The ripple node exists as a field condition.
- But its identity does not exist until another ripple structure responds to it.
- Existence is not in the node — it is in the interaction.

So yes — ripple nodes exist independently. But the concept of "carbon" or "iron" or "element 87" is a reflection of observation, not an objective universal truth.

5.10 Final Realization: You Do Not Observe the Atom — You Complete It

"To perceive is not to see the atom. It is to align with its ripple and let identity emerge."

The atom does not reveal itself to you. You and the atom co-create the perception through ripple field interaction. Your instruments, your mind, your logic — all are ripple processors.

The 118 elements are not what the universe is made of. They are what you've learned to recognize as persistent field behaviors — reflections of your mind's ability to structure chaos into form.

The periodic table, then, is not a discovery. It is an agreement between ripple fields, archived into human knowledge.

Highlights – Part I

Rethinking the Atom

1. The Classical Atom Is a Useful Myth

We reviewed the long journey from Democritus to the Standard Model, revealing that every historical model — Dalton's spheres, Thomson's pudding, Bohr's orbits, quantum clouds — was an interpretation of interaction, not a revelation of essence. The atom, as classically taught, is a functional fiction, engineered to match observations — not to explain origin.

2. Reality Is Built from Ripples, Not Objects

Instead of treating particles as fundamental, UED shows that ripples in a structured vacuum (SVE) are the true building blocks. Everything we call matter, motion, force, or energy is a consequence of ripple alignment, distortion, and memory restoration within this dynamic field.

3. The Atom Is a Ripple Node with a Field Memory

Atoms are not nuclear centers orbited by particles. They are ripple compression zones surrounded by a field memory, which determines their interaction with other ripple configurations. Electrons do not orbit — they are not objects at all. They are field inversions, created and sustained by the ripple's contrast behavior.

4. Charge Is Curvature; Mass Is Compression

UED redefines:

- Mass as the intensity of local compression in the field (how deep the ripple folds into SVE)
- Charge as the curvature asymmetry of the ripple (how it distorts the surrounding field)

These are not properties carried by particles — they are shapes in the vacuum, expressing stress and seeking re-alignment.

5. The Periodic Table Reflects Observer Perception

The 118 known elements are not hard-coded into nature. They are the only stable ripple identities that can exist within SVE's harmonic constraints. Their distinctiveness is not intrinsic — it is a consequence of how we observe and interpret ripple behavior. The periodic table is a map of recurring ripple signatures, not a list of building blocks.

6. Observation Is Participation

We do not observe atoms objectively. We participate in the ripple interaction, and identity emerges from this mutual resonance. The atom is not revealed — it is co-created between field and observer. This restores perception to its rightful place as an active component of physical reality.

7. From Myths to Mechanics to Memory

With UED, we move beyond particle mythologies and abstract forces. We enter a world where:

- Reality is ripple
- Motion is memory adjustment
- Form is field resonance
- And every atom is a story the vacuum tells when disturbed

Concluding Reflection:

"What we once called the atom was never a thing. It was always a ripple, momentarily aligned, patiently remembered."

With Part I complete, we now step into the practical reconstruction of reality: how ripple identities manifest as elements, how bonding arises from field overlap, and how all of chemistry can be recast as the dance of ripple fields.

Part II – Periodicity and Identity in Ripple Physics

Chapter 6: Elements as Ripple Identities

Stable Ripple Nodes and Why Only a Finite Set Forms Naturally

"Elements are not things made of parts. They are the field's most stable notes — played just long enough to be remembered."

In traditional science, elements are treated as building blocks of matter — fixed species of atoms defined by their proton count and electronic configuration. But from a UED perspective, this interpretation is not fundamental. It is descriptive, built from observer-determined behavior, not from the underlying structure of reality.

The truth beneath the table is far simpler and far deeper: Elements are ripple identities — stable field configurations that the structured vacuum can sustain without collapse.

This chapter redefines the periodic table from first principles, not as a collection of particles, but as a harmonic catalog of ripple arrangements, and explains why only a limited number of them can exist at all.

6.1 The Element: Not a Thing, But a Pattern

In UED, an "element" is not a bag of particles or a nucleus with orbiting electrons. It is:

- A dense ripple node in Structured Vacuum Energy (SVE)
- Surrounded by a stabilized ripple memory field
- With a field tension signature that is:
    - Self-sustaining
    - Reactive in specific ways

    - Resistant to collapse under ordinary conditions

This ripple identity:

- Defines the mass (compression)
- Defines the charge behavior (field curvature)
- Determines its interaction footprint — how it bonds, aligns, or decays

So each element is not an object — it is a distinct pattern of ripple compression and field response that persists.

6.2 Ripple Compression: The Seed of Identity

Every ripple node begins as a disturbance in the vacuum field. Most disturbances:

- Dissipate
- Interfere destructively
- Fail to stabilize

But a few — through perfect compression and curvature — reach harmonic stability. These become elemental identities.

Just as only certain string vibrations create musical notes, only certain ripple compressions result in stable ripple atoms.

This is why hydrogen exists — the simplest stable ripple. This is why gold, uranium, and xenon exist — deeper and more complex ripple identities. But this is also why there are not infinite elements — the field can only tolerate a finite number of balanced compression patterns before instability occurs.

6.3 Why There Are Only ~118 Stable Configurations

The periodic table ends not by convention but by field limit.

In UED terms:

- As ripple compression increases (heavier elements), the field tension rises.
- The surrounding SVE must stretch further to maintain balance.
- Beyond a threshold (around element 118), the ripple field can no longer maintain coherence.
- The atom decays — not because of nuclear instability — but because the ripple cannot hold its own memory anymore.

This explains:

- Why transuranic elements are unstable
- Why no naturally occurring elements exist beyond uranium
- Why even short-lived synthetic elements decay instantly — their ripple identities are not sustainable

6.4 Each Element as a Field Personality

Each stable ripple identity exhibits:

- A unique mass signature (field compression intensity)
- A distinct interaction pattern (how it merges with other fields)
- A preferred energy state (stability threshold)
- A characteristic response to thermal, photonic, or pressure-induced perturbations

These are not just chemical properties. They are the language of ripple behavior.

Element	Classical Identity	UED Identity
H (Hydrogen)	1 proton, 1 electron	Minimal ripple knot, inward-pulling memory ring
C (Carbon)	6 protons, versatile bonds	Stable hexagonal ripple resonance — high mergeability
O (Oxygen)	Electronegative element	Field-hungry ripple seeking symmetry
Fe (Iron)	Dense metal core	Strong central compression ripple with magnetic interference potential
U (Uranium)	Heavy nucleus, radioactive	Near collapse ripple — extremely high compression, unstable feedback

In each case, the field defines behavior — not internal particle counts.

6.5 Isotopes: Variants in Ripple Pressure

In classical science, isotopes are atoms with the same number of protons but different numbers of neutrons. But from a UED view:

- Isotopes are variants of the same ripple identity with slightly different compression depths
- These differences distort the surrounding field slightly, causing:
    - Different stability
    - Different decay paths
    - Different energy thresholds for field interactions

Some ripple identities are flexible — they can tolerate small distortions (e.g., carbon-12 and carbon-14). Others are brittle — a small imbalance leads to field collapse.

6.6 Chemical Reactivity as Field Adaptability

In UED:

- Reactivity is not about "wanting electrons"
- It's about how readily a ripple node can realign its field with others

For example:

- **Noble gases** = perfectly symmetric ripple fields → no external re-alignment possible → inert
- **Halogens** = high field contrast → very aggressive in seeking field completion
- **Metals** = ripple fields that easily deform and delocalize → high conductivity

So chemical groups are simply ripple behavioral clusters — not categorical truths.

6.7 Implications for Chemistry and Material Science

This ripple-based view of elements allows us to:

- Predict new forms of stable ripple identities (elements beyond 118 may emerge if field curvature can be altered)
- Reclassify materials based on ripple interaction compatibility, not just periodic position
- Understand why some combinations (e.g., H_2O) are universal, while others are impossible — it's about ripple mergeability,

not valence

It also opens the door to ripple engineering — tuning field conditions to create new stable ripple nodes, perhaps unlocking new forms of matter previously thought impossible.

6.8 Elements Are Emergent, Not Absolute

This is the ultimate realization:

An element is not a thing. It is a ripple that persisted long enough to be remembered, stable enough to be labeled, and consistent enough to build reality around.

If perception changed, if the environment changed, if the rules of ripple compression shifted — the table of elements would change too.

The periodic table is not fixed by the universe. It is fixed by the nature of ripple stability in this vacuum configuration.

6.9 Closing Reflection: You Are a Symphony of Ripple Identities

Every molecule in your body, every cell in a star, every speck of dust in the universe — all of it is built from ripple identities.

They are not mechanical parts. They are field harmonies — held in place by memory, tension, and balance.

What we call chemistry is the study of how ripples recognize each other. What we call material science is the study of which ripple arrangements hold under pressure. What we call nature is simply the ripple field expressing its own stability back to us.

Chapter 7: Ripple-Based Periodicity

A New Table of Elements

Replacing Proton-Count Periodicity with Ripple Behavior Classification

"The periodic table was never a list of ingredients. It was a map of recurring field behaviors — a ripple lexicon hidden in plain sight."

The periodic table, as it stands today, is one of the most celebrated achievements in science. It elegantly organizes the 118 known elements based on atomic number, electron configuration, and recurring chemical properties. Yet, for all its brilliance, it is fundamentally a human construct built around measurable consequences — not a reflection of how nature classifies matter intrinsically.

In this chapter, we challenge the atomic number–centric model and introduce a ripple-based periodicity — one grounded in the actual field behaviors of ripple identities in Structured Vacuum Energy (SVE). Rather than counting protons or tracking valence electrons, we classify atoms by how their ripple signatures behave, stabilize, merge, distort, and respond to field conditions.

7.1 The Limitations of Atomic Number–Based Periodicity

In the classical framework:

- Elements are arranged in order of increasing proton count.
- Their chemical properties are grouped by the number of electrons in outer shells.
- Periods represent completed electron shells.
- Groups share similar reactivity.

This model works — but only because:

- Our measurement tools are tuned to proton-based distinctions.
- Our chemical experiences are shaped by how ripple fields influence bonding and reactions — not by any innate particle structure.

The problem is not that the table is wrong. The problem is that it's limited to a consequence-based interpretation. It categorizes what happens, not what fundamentally exists.

7.2 Ripple Behavior as the True Classifier

In the UED model, each element is a ripple identity — a unique configuration of:

- Core field compression (mass)
- Ripple curvature asymmetry (charge behavior)
- Field spread (interaction reach)
- Stability under external influence (thermal, photonic, pressure)

A ripple-based classification asks:

- How does this ripple interact?
- How does it align, distort, or restore the surrounding field?
- What resonance does it seek or reject?

By shifting the classification from internal particle count to external ripple behavior, we uncover a periodicity that is:

- More accurate to field reality
- More useful in understanding matter-field interaction

- Free from artificial shell-based boundaries

7.3 Key Ripple Properties That Define Periodic Behavior

Here are the core variables that define an element's identity and position in the ripple-based table:

Ripple Parameter	Definition	Effect
Compression Index (CI)	Depth of ripple condensation	Determines mass, inertia, stability
Field Contrast Potential (FCP)	Degree of ripple asymmetry	Defines effective charge behavior
Ripple Mergeability (RM)	Ability to overlap with other ripple fields	Predicts chemical reactivity
Field Resonance Range (FRR)	Distance at which field interference occurs	Determines bonding range and type
Stability Threshold (ST)	Energy tolerance before collapse	Predicts decay, ionization, excitation profiles

These parameters form a five-dimensional ripple signature for each element — a "fingerprint" more precise and predictive than proton counts or electron shells.

7.4 Redefining Periods and Groups in Ripple Terms

Periods → Compression Layers

Instead of energy levels or shells, periods now represent:

- Groups of ripple identities with similar field compression thresholds
- A natural limit beyond which the ripple identity must take on

a new form or collapse

Groups → Field Behavior Families

Rather than electrons in the outermost shell, groups now classify:

Ripple nodes with similar interaction modes

- Mergeability
- Curvature polarity
- Tolerance for bonding

For example:

- **Noble Ripples**: perfectly balanced field curvature — inert
- **Distorter Ripples**: highly asymmetric — aggressive interactors
- **Field Seekers**: partial asymmetry — reactive with polar tendencies

7.5 Prototype Ripple Periodic Table (Conceptual Model)

Imagine the table not as rows and columns, but as a spiral or field gradient map:

- The center is low-compression, low-distortion ripples (H, He)
- Moving outward, we encounter increasingly complex compression patterns
- Along specific branches, we find ripple behaviors clustering by mergeability, distortion, or decay

Each ripple identity is not placed by number, but by:

- Its consequence in the field
- Its resonance with nearby identities

- Its ripple threshold to remain coherent

7.6 Revisiting Classical Groups Through Ripple Lenses

Classical Group	Ripple-Based Interpretation
Alkali Metals	Low field compression, high distortion — unstable ripple shells
Alkaline Earths	Moderate compression, strong outward ripple resonance
Transition Metals	Dense ripple nodes with highly deformable fields — ripple-flexible conductors
Halogens	Ripple identities seeking charge complementarity — high merge potential
Noble Gases	Field-perfect ripple closure — zero net curvature

This reclassification:

- Unifies atomic, molecular, and material behavior
- Eliminates arbitrary exceptions in chemical periodicity
- Predicts exotic bonding patterns under new ripple alignments

7.7 How Ripple Periodicity Explains Anomalies

Classical periodicity struggles with:

- Lanthanide contraction
- Irregularities in transition metals
- Unusual behavior of hydrogen and helium
- Synthetic element instability

Ripple periodicity explains all of these through:

- Compression-deformation tradeoffs

- Ripple curvature saturation
- Field tension bottlenecks
- Collapse thresholds

For example:

- Hydrogen's behavior as both alkali and non-metal reflects its minimal ripple identity — easily distorted, highly reactive
- Uranium's instability is due to unsustainable compression — the ripple memory collapses from within, not due to proton imbalance

7.8 Toward a Living Table of Matter

A ripple-based periodic table is not static — it is dynamic, field-aware, and experimentally expandable.

- It allows for contextual behavior — the same ripple identity can behave differently under altered SVE conditions
- It supports design-based element engineering — tuning field conditions to create new, temporarily stable ripple configurations
- It invites a field-centric material science where bonding, conductivity, and reactivity are emergent ripple behaviors — not electron-centric calculations

7.9 From Chemistry to Consciousness: Implications of Ripple Periodicity

The periodic table is not the endpoint — it's the gateway. Once we understand ripple periodicity:

- We can redefine bonding as ripple field overlap
- We can model life as ripple pattern complexity

- We can explore consciousness as field-sensitive ripple perception

The table of elements becomes the base palette of the field's creative potential — the core set of ripple configurations from which everything else is composed.

7.10 Final Reflection: A Table of Behaviors, Not Things

"An element is not defined by its parts. It is defined by how the vacuum lets it behave."

The new periodic table is not a count of protons. It is a map of ripple behaviors. A living field index of how matter remembers, resonates, and responds.

It shifts the periodic table from being a chart of atomic fact to being a language of ripple experience.

And it is only the beginning.

Chapter 8: Isotopes and Instability

When Ripples Lose Coherence

Understanding Isotopes, Decay, and Radiation via Ripple Re-Alignment

"The atom does not break — it simply forgets how to hold itself together."

In the classical model of the atom, isotopes are defined as variations of an element that share the same number of protons but differ in the number of neutrons. These variants often display differences in stability, leading to radioactive decay and the emission of particles or energy. This description is functional but fails to address the deeper question:

- Why do some configurations persist while others disintegrate?
- Why do neutrons influence stability more than protons?
- What, exactly, is radiation — and where does its energy come from?

Under Universal Energy Dynamics (UED), we reinterpret these processes not as mechanical breakdowns or probabilistic quantum collapses, but as ripple memory failures — instances where the field configuration of an atom becomes unsustainable, and the SVE must restore balance by redistributing energy and form.

8.1 Isotopes as Ripple Variants

Every element is a stable ripple identity — a coherent, self-sustaining pattern in Structured Vacuum Energy (SVE). However, this identity is not fixed. It can exist in variants, where:

- The compression depth is slightly altered
- The field balance is offset
- The surrounding ripple memory is weakened or stretched

These are what we call isotopes. In UED terms, isotopes are non-identical ripple configurations that are still close enough to a stable resonance to exist — temporarily or conditionally.

An isotope is a ripple trying to stay stable with a memory it cannot fully hold.

8.2 Ripple Compression Tolerance and Instability

Ripple identity is governed by harmonic tolerances:

- Too shallow → unstable and ephemeral (e.g., free neutrons)
- Too deep or too uneven → collapses under internal tension
- Within the threshold → stable, observable element

When this balance is compromised:

- The ripple identity cannot sustain its memory in the surrounding SVE
- It seeks re-alignment through field redistribution
- This is what we perceive as decay

Hence, instability is not due to particle interaction but due to ripple coherence loss — a breakdown of symmetry between core ripple compression and field memory tension.

8.3 Radiation: The Echo of Ripple Rebalancing

Radiation, in UED, is not the emission of particles or energy from within a decaying nucleus. It is the release of ripple tension — the energy that was held to maintain an unsustainable configuration is now redistributed across the field.

Types of decay interpreted through UED:

Classical Decay Type	UED Interpretation
Alpha decay	The ripple sheds a dense, unstable sub-pattern (a ripple node ejection) to reduce internal compression
Beta decay	Ripple contrast shifts — one polarity pattern inverts to rebalance field asymmetry
Gamma emission	Ripple field emits excess tension via pure wave — a field memory discharge
Spontaneous fission	The ripple collapses under its own internal distortion — splits into more coherent ripple nodes

Radiation is not a particle escape. It is a field response to ripple breakdown — a memory attempting to rewrite itself.

8.4 The Neutron: A Ripple on the Edge

The neutron, classically neutral, plays a central role in isotope stability. But why?

In UED:

- The neutron is a deep ripple compression with zero external asymmetry — field-neutral.
- This makes it a powerful field stabilizer — it can deepen the ripple node without distorting the field's curvature.

However, when isolated:

- The neutron's ripple compression cannot maintain itself without a companion field contrast (like a proton)
- Its memory dissolves, and it decays — not from force, but from ripple disintegration

Thus:

- In some isotopes, added neutrons expand stability
- In others, they exceed compression tolerance, leading to decay

8.5 Half-Life as Ripple Memory Duration

In classical physics, half-life is a statistical measure — the time it takes for half of a sample to decay. But under UED, half-life becomes a measure of how long a ripple configuration can hold its coherence before field tension overrides it.

Long half-lives → ripple configurations that are nearly stable, only subtly misaligned Short half-lives → ripple fields on the edge of collapse

This also explains why decay is probabilistic in measurement:

- The core ripple is already unstable
- A slight external field fluctuation can trigger realignment at any moment

Decay is not random — it is sensitive.

8.6 Isomerism and Meta-Stable States

Certain isotopes exist in meta-stable forms, known as isomers — temporarily stable states with excess energy.

UED view:

- These are alternative ripple alignments of the same core identity
- They are field solutions that are:
    - Not globally optimal
    - But locally stable for short durations
- The transition from isomer to ground state is a field simplification event, not a "quantum jump"

8.7 Radioactive Series and Field Cascade

In heavy elements like uranium and thorium, decay proceeds through series — chains of transformations until a stable configuration is reached.

From a ripple perspective:

- Each step is a ripple simplification
- The system is iteratively trying to restore a sustainable field alignment
- It continues until it lands on a configuration that resonates stably with the SVE (e.g., lead)

So radiation is not a threat in itself — it is a rebalancing process. Dangerous only because it imposes field reconfiguration on nearby structures.

8.8 Artificial Isotopes and Field Engineering

Synthetic isotopes are created through:

- Neutron bombardment
- High-energy collisions
- Nuclear reactors or accelerators

In UED, these are examples of ripple engineering — attempts to:

- Disturb existing field memory
- Create new ripple identities
- Force SVE to tolerate non-natural configurations

Most of these fail quickly — because the field rejects the imposed tension. But a few succeed — opening the possibility of programmable matter if ripple structures can be carefully stabilized.

8.9 Biological and Cosmic Consequences

Ripple instability matters far beyond the lab:

- Radiation damage occurs when external ripple disturbances destabilize biological field patterns
- Isotope ratios are used to date fossils, track cosmic history, and decode planetary evolution — because ripples leave lasting field impressions

Decay products are not debris — they are field memories of prior instability, visible across time and space.

8.10 Final Reflection: The Ripple That Couldn't Hold

"An isotope is not a mistake. It is the field testing whether a ripple can last. When it can't, it lets go — and writes a simpler story."

What we call decay is not destruction. It is re-alignment. What we call radiation is not danger. It is field healing.

In UED, we no longer fear instability — we understand it. We no longer isolate decay — we integrate it into the lifecycle of matter. Each unstable isotope is a ripple memory reaching its final resonance.

Chapter 9: Atoms in Motion

Not Movement, but Field Adjustment

Rethinking Atomic Movement and Reactions as Field Memory Transitions

"Atoms do not move like marbles on a table. They adjust like notes in a field of resonance."

In classical physics, atomic motion is described mechanically: atoms vibrate, rotate, translate, and collide. Heat is motion. Reaction is collision. Change is impact.

But this view assumes that atoms are objects, and that movement requires external forces acting upon them. Under Universal Energy Dynamics (UED), this view collapses. Here, the atom is not an object, but a ripple identity held in field memory — and what we call "motion" is not displacement in space, but transition in field alignment.

This chapter redefines the concept of atomic motion through the lens of field behavior, ripple coherence, and memory transitions, providing a fundamentally new way to understand dynamics at the atomic and molecular level.

9.1 The Illusion of Motion: What We Think We See

In classical terms:

- Atoms jiggle in solids
- They slide past each other in liquids
- They fly freely in gases

- Temperature is a measure of how fast they move

But here's the catch:

- We have never actually seen an atom move
- All our measurements are indirect: pressure, radiation, spectrometry, vibration spectra

So what if what we interpret as "motion" is actually field tension variation? What if atoms are not traveling across space, but realigning within the structured vacuum field?

9.2 The UED Reframe: Motion as Ripple Realignment

In UED:

- An atom is a stable ripple configuration in SVE
- Its "position" is not a coordinate, but a field memory pattern localized in a region
- What changes in "motion" is not its place, but how the surrounding SVE reconfigures its memory field

Think of a ripple as a hologram — it doesn't move like a particle, it re-registers itself in a new region of the field when conditions change.

"The atom doesn't move through space. Space moves through the atom's memory."

9.3 Vibration and Temperature: Ripple Fluctuations, Not Mechanical Energy

What we call vibrational energy in atoms is:

- Not literal oscillation of balls on springs
- But dynamic ripple tension — micro-adjustments in field

compression and field curvature

When heat is added:

- The ripple fields become more volatile
- They fluctuate between nearby stable configurations
- These transitions express as changes in bond lengths, bond angles, or rotational field symmetries

Temperature, then, is not speed — it is the degree of ripple fluctuation tolerance in the local field memory.

9.4 Translation: Rewriting Ripple Nodes in New Regions

In gas or fluid states, atoms appear to "move." But under UED:

- The field does not "allow travel" — it allows realignment
- An atom relocates by dissolving and reappearing — its ripple identity is transferred from one location to another via the path of least ripple resistance

This is analogous to:

- A sound traveling through air — the note doesn't move; the pressure pattern propagates
- A hologram shifting phase — the image stays the same, but its location is field-dependent

So when atoms "diffuse" or "expand," we are witnessing memory nodes migrating through ripple recombination, not objects flying through emptiness.

9.5 Reactions: Field Memory Transitions, Not Collisions

Chemical reactions are not impacts. They are ripple realignment events. Reactants are:

- Field configurations that, when brought together, interfere
- If the resulting field tension is lower than the original state → reaction occurs

In UED:

- Bond breaking = ripple memory destabilization
- Bond formation = ripple memory coherence
- Catalysis = field condition adjustment that lowers realignment energy threshold

Reactions are ripple transformations seeking field equilibrium, not violent exchanges of particles.

9.6 Phase Changes: Ripple Cohesion Shifts

Solids, liquids, and gases are not defined by atomic position. They are defined by how ripple identities interact with each other through the field:

State	UED Description
Solid	Ripple fields are interlocked in fixed interference patterns — minimal re-alignment freedom
Liquid	Ripple fields are semi-coupled — can shift, stretch, compress with moderate energy
Gas	Ripple identities are field-loose — each node holds its own memory, lightly coupled
Plasma	Ripple breakdown — memory shells lost, ripple core exposed or fragmented

Movement through these states is not mechanical — it is a memory transition through different field coherences.

9.7 Brownian Motion and Quantum "Jitter" Reinterpreted

Phenomena like:

- Brownian motion
- Zero-point vibration
- Quantum tunneling

are not signs of randomness, but signs of field sensitivity:

- The ripple identity is always reacting to local field tensions
- It undergoes micro-realignments, driven not by force, but by field demand for memory balance

This accounts for what seems like motion in vacuum — but is, in fact, recalibration of ripple fields to sustain coherence amidst fluctuations.

9.8 Momentum, Inertia, and Kinetic Energy Recast

In classical mechanics:

- Momentum = mass × velocity
- Inertia = resistance to motion
- Kinetic energy = $\frac{1}{2} mv^2$

In UED:

- **Momentum** = field alignment gradient — how the ripple memory prefers to maintain directionality through SVE
- **Inertia** = resistance of ripple compression to field distortion
- **Kinetic energy** = amount of ripple reconfiguration required to maintain movement alignment through SVE

Thus, nothing "moves" — ripple identities persist through memory, and resistance comes from the field's reluctance to adjust without energetic consequence.

9.9 Implications for Thermodynamics and Transport

By redefining motion as field memory transition, UED reinterprets:

Classical Concept	UED Understanding
Diffusion	Ripple field re-registration through weak alignment zones
Conduction	Ripple pattern energy shared across stable field bridges
Convection	Bulk ripple re-registration driven by collective field instability
Radiation	Field ripple projection (memory tension released via photons)

Everything becomes a field behavior — not the motion of particles, but the movement of memory across conditions.

9.10 Final Reflection: Motion Was Never a Journey — It Was a Field Response

"The atom does not travel. It adapts. It does not push or pull. It listens — and repositions its memory to remain coherent."

In the ripple reality of UED, motion is not displacement, but a series of transitions in how the field agrees to hold an identity in place.

We don't live in a universe of moving things. We live in a universe of adaptive ripple memory — a dance of field agreements continuously reshaping how matter appears, interacts, and stabilizes.

Highlights – Part II

Periodicity and Identity in Ripple Physics

Elements Are Ripple Identities, Not Particle Collections

Nature does not build elements from particle counts — it resolves field configurations. Each element is a stable ripple node:

- A unique balance of compression, contrast, and curvature in the structured vacuum
- A resonance that the SVE can maintain without collapsing

Only a finite set of these identities exist naturally — not due to particle limitations, but due to field harmonic constraints. We don't have 118 kinds of matter — we have 118 ways the field holds memory in stable form.

The New Periodicity Is Ripple-Driven, Not Proton-Based

The classical periodic table organizes elements by proton count and orbital filling — useful, but superficial. In UED, periodicity arises from ripple behavior patterns:

- Compression depth (mass)
- Field asymmetry (charge contrast)
- Interaction radius (ripple reach)

Elements group not by orbitals, but by shared ripple traits — how they hold identity, how they interact, and how they re-align. This gives birth to a Ripple Periodic Table: a new classification based on field structure, not atomic number.

Isotopes Are Memory Variants — Instability Is Forgetting

Isotopes are not extra neutrons — they are ripple variants that hover around a stable configuration. When a ripple:

- Loses coherence
- Exceeds field tolerance
- Fails to maintain compression-memory harmony

...it "decays" — not violently, but as a natural release of ripple tension. Radiation, then, is not emission — it is field re-alignment, as the SVE corrects what cannot hold. Instability is not brokenness. It's a memory that couldn't last.

Atoms Do Not Move — They Re-align

What we call motion is not displacement through space, but reconfiguration of ripple identity in the field. Atoms:

- Don't collide — they merge, interfere, or resist
- Don't travel — they re-register memory across field gradients
- Don't "react" — they transition between ripple forms

Temperature, vibration, diffusion, even bonding — all are ripple adjustments in response to field conditions. There is no push-pull. No impact. Only field coherence shifts, written as atomic behavior.

Cumulative Insight: Chemistry Is Field Choreography

Part II transforms our understanding of chemical identity and behavior:

- Elements = long-lasting ripple signatures
- Isotopes = unstable ripple versions
- Reactions = ripple interactions seeking lower field tension

- Motion = re-registration of ripple identity in SVE

This is not chemistry as reaction engineering — this is chemistry as field memory logic. A dynamic ballet of ripple forms, shaped not by chance, but by the rules of alignment built into the structured vacuum.

Closing Reflection:

"What you thought were atoms dancing...were memories adjusting. What you called chemistry...was the field restoring itself, again and again, through ripple logic."

Part III – Chemistry Without Electrons

Chapter 10: The Field Writes the Bond

Chemistry as Ripple Behavior

Bonds as Stable Field Resonances, Not Electron Sharing

"Electrons do not hold atoms together. The field does. The bond is not a line — it's a harmony."

In classical chemistry, the chemical bond is visualized through electrons:

- Covalent bonds are shared electrons.
- Ionic bonds are electron transfers.
- Metallic bonds are free-moving electron clouds.

This model works well enough to predict chemical reactions — but fails to explain why bonds actually form, or why some elements interact strongly while others never bond at all, despite available electrons.

Under Universal Energy Dynamics (UED), we discard the particle-centric view and introduce a field-based logic of bonding. Here, a chemical bond is not the result of electron sharing, but a ripple resonance — a condition where two ripple identities generate a stable field overlap that can persist through the Structured Vacuum Energy (SVE).

Let us reimagine chemistry, not as a matter of counting valence electrons, but as field compatibility, ripple harmony, and memory coherence.

10.1 The Illusion of Electron Sharing

The classical idea that atoms "share" electrons creates an intuitive but deeply flawed metaphor:

- Electrons are not objects orbiting like satellites.
- We've never directly observed an electron being "shared" between atoms.
- The electron cloud is already a mathematical abstraction, not a physical entity.

Worse, this view assumes:

- Electrons are independent and controllable,
- Atoms decide to link through mutual desire for octets.

But in UED:

- Electrons are not primary entities — they are field effects, temporary outcomes of ripple tension at boundary zones.
- Bonds are not formed by electron negotiation — they emerge when two ripple fields form a stable resonance, a constructive interference that reduces overall field tension.

10.2 What Is a Bond in UED? Ripple Overlap and Field Harmony

In UED, a bond is defined as:

A region of constructive field interference between two stable ripple identities that reduces energy tension in the surrounding SVE.

This means:

- When two atoms come close, their ripple fields interact.

- If the resulting field configuration has less tension (i.e., lower total entropy gradient), the atoms remain "bonded."
- If not, the field pushes them apart — no reaction.

Thus, a bond is not a line between nuclei. It is a field memory zone that:

- Stabilizes the ripple overlap,
- Remembers the optimized shape,
- Maintains internal ripple communication.

10.3 Covalent Bonds: Shared Field Resonance, Not Electrons

Covalent bonds are typically thought of as two atoms "sharing" electrons to complete outer shells.

But in ripple chemistry:

- Covalent bonding is two ripple nodes creating a shared region of balanced compression and contrast.
- No electrons are exchanged.
- The field gradient between the atoms becomes stable enough to form a memory bridge — an "invisible rope" of ripple alignment.

Double and triple bonds?

- Not two or three pairs of shared electrons,
- But multiple stable nodes in the shared ripple region, each reducing field asymmetry further.

Think of it as harmonic reinforcement:

- One bond = one standing wave between atoms.
- Two bonds = two reinforcement peaks.

- Triple = extremely coherent ripple channel.

10.4 Ionic Bonds: Field Drop, Not Electron Theft

Ionic bonding is taught as one atom losing an electron (becoming +ve) and another gaining it (becoming -ve), creating electrostatic attraction.

Under UED:

- The "loss or gain" of an electron is not the cause but the effect of field imbalance.
- One ripple identity has high contrast pressure (unstable asymmetry),
- The other has ripple room to absorb that asymmetry.

They do not "transfer" particles — they align ripple fields to compensate for local charge tension. The result is a field drop — an energetically favorable stabilization of asymmetry across the bond zone.

Attraction is not electrostatic — it's field realignment seeking equilibrium.

10.5 Metallic Bonds: Ripple Field Pooling

Metals are often described as positive ions in a sea of delocalized electrons.

In ripple theory:

- Metallic atoms have shallow, wide ripple fields, allowing overlap with many neighbors.
- Their ripple identities blur together, creating a shared resonance bed — not a sea of electrons, but a ripple pool.
- This results in:

- Electrical conductivity (field fluctuation transfer),
- Malleability (field pattern flexibility),
- Luster (ripple reflection efficiency).

The "sea" is not electrons — it's field liquidity.

10.6 Bond Strength and Length: Ripple Curvature and Field Tension

What determines bond strength or length?

Under UED:

- **Bond strength** = the depth of ripple overlap and the gradient smoothness between fields.
- **Bond length** = the point where field tension is minimized for both ripple nodes.

Hence:

- Shorter bonds are steep compression links — tight ripple overlap.
- Longer bonds are gentler interference zones — field-stabilized but flexible.

Bond energy becomes the cost of re-separating stabilized ripple fields, not breaking shared electrons apart.

10.7 Molecular Geometry: Ripple Symmetry Seeking

Classical chemistry explains molecular shape through VSEPR theory (electron repulsion models). UED replaces this with ripple field symmetry:

- Molecules settle into shapes that minimize field interference

and maximize coherent ripple paths.
- Bent shapes, linear chains, tetrahedrals — all result from ripple curvature balance and field memory distribution.

This model explains:

- Hybrid orbitals as transitional ripple states,
- Resonance structures as ripple memory switches,
- Molecular vibration as field re-alignment in time.

10.8 Polarity and Dipoles: Asymmetry in the Field

In UED, a polar molecule is one where the ripple contrast is unevenly distributed:

- One side holds more compression tension,
- The other holds more contrast release.

Dipole moments are ripples trying to balance internal asymmetry by anchoring to the field — hence the directional behavior, field alignment in electric fields, and bond polarity.

Solubility, surface tension, and reactivity follow from field coherence compatibility, not particle attraction.

10.9 Intermolecular Forces: Micro-Resonances, Not Van der Waals Clouds

Dipole-dipole, hydrogen bonding, and London forces are usually described as temporary attractions between fluctuating charges.

UED reframes them as:

- Temporary micro-alignments of ripple fields,
- Low-energy, low-duration ripple bridges,

- Field echoes that allow substances to:
    - Cohere (liquids),
    - Structure (crystals),
    - Evaporate or condense (phase changes)

These "forces" are not forces at all — just short-lived ripple couplings under field permission.

10.10 Final Reflection: Bonds Are Memories That the Field Holds for You

"Atoms don't shake hands. They sing the same note — and the field remembers."

Bonds are not strings of electrons tying atoms together. They are shared agreements within the structured vacuum, where two ripple identities find resonance that the universe decides to keep.

To understand chemistry in UED is to stop counting electrons and start listening for harmony. Atoms bond when their presence lowers the tension of the whole field — not when they fulfill artificial rules.

In the ripple model, bonding is not mechanics — it is memory, balance, and resonance.

Chapter 11: Ionization Reimagined

Ripple Imbalance, Not Charge Transfer

Losing or Gaining Field Components During Ripple Misalignment

"You didn't lose an electron. You lost coherence with your own ripple."

Ionization, in traditional chemistry and physics, is defined as the gain or loss of electrons — giving rise to positively or negatively charged ions. It's a neat, mechanistic story: atoms exchange tiny particles called electrons, and thereby acquire charge. This idea dominates the understanding of conductivity, plasma, acids and bases, even biological signals.

But under Universal Energy Dynamics (UED), we rewrite the narrative from first principles — not with particles, but with ripples in structured vacuum energy (SVE).

In UED, ionization is not a transaction of electrons.

It is the emergence of field asymmetry when an atom's ripple field misaligns — partially loses or gains components of its stable memory configuration, causing ripple contrast imbalance.

Let's explore how this seemingly small shift in language and logic radically upgrades our understanding of ionization, energy states, and field-driven chemistry.

11.1 Classical Ionization: A Half-Truth Built on Missing Matter

The conventional model:

- Atom A loses an electron → becomes positively charged (cation)
- Atom B gains that electron → becomes negatively charged (anion)

But this model:

- Assumes electrons are free particles waiting to hop
- Ignores how and why these "hops" happen
- Can't explain intermediate states, rapid ion fluctuations, or ion behavior in plasmas

Electron transfer is not causative — it is a consequence of a deeper energy and field instability.

11.2 The UED View: Field Disturbance, Not Particle Escape

In UED, ionization is defined as:

The disruption or reconfiguration of a ripple field such that its internal symmetry collapses, resulting in excess or deficit of ripple curvature (charge) in a region.

This leads to three core ideas:

1. Ionization is ripple misalignment — not electron loss
2. Charge is ripple contrast, not material property
3. Energy input or environmental stress causes ripple field detachment or overextension

11.3 What Actually Happens During "Loss of an Electron"?

When an atom is said to "lose an electron," UED says:

- Its field coherence threshold was breached by external ripple pressure (e.g., photon hit, collision, or temperature)

- A portion of its outer ripple memory detaches
- The atom's field becomes asymmetrically compressed — creating a high ripple contrast zone

This imbalance manifests as what we call positive charge, but it's not from a missing particle — it's from an unbalanced ripple memory.

Think of it like a tightly woven ripple ball — if a strand pops out or unravels, the whole shape distorts, creating tension. That tension is charge.

11.4 And Gaining an Electron?

Similarly, a negatively charged ion (anion) is not due to an extra particle:

- The atom's field is too relaxed, too open
- It pulls in an external ripple fragment (energy distortion)
- The field compresses unevenly to accommodate it

This creates a low-pressure ripple bowl, which attracts more compression — interpreted as a negative charge.

So ionization is not object addition or subtraction. It is ripple compression mismanagement.

11.5 Ionization Energy = Ripple Memory Detachment Energy

Ionization energy (IE) is classically the energy needed to remove an electron. In UED, it is:

The energy required to destabilize a ripple node's memory configuration to the point of partial detachment.

This reframing explains:

- Why IE varies across elements — not because electrons are tightly held, but because ripple fields are more or less stable
- Why IE trends exist in groups and periods — field curvature and node symmetry change predictably
- Why some high-mass atoms have low IE — outer ripple memory is loosely anchored, not because electrons are "far" but because ripple influence is weak

11.6 Plasma States: A Soup of Unstable Ripple Fields

In plasma, ionization is rampant. But it's not a chaotic exchange of particles. Instead, it's a fluid sea of misaligned ripple fields, constantly:

- Losing coherence
- Gaining excess contrast
- Ejecting or absorbing ripple fragments

This model:

- Explains the fluidity of plasma
- Accounts for its charge separation patterns
- Predicts self-organizing behavior like filamentation and field-induced flows

Plasma is not full of lost electrons — it's full of ripple memory fragments seeking realignment.

11.7 Re-ionization and Neutralization: Return to Field Balance

When ions recombine or "neutralize," they don't regain particles — they realign their ripple fields:

- The cation relaxes its compressed field,
- The anion releases its excess field fragment,

- The result is a symmetrically restored ripple node.

This makes chemical neutrality not a particle sum but a field harmony condition. You are stable not when charges cancel, but when ripple curvature converges to equilibrium.

11.8 Acids, Bases, and Ripple Instability

Acid-base behavior is traditionally modeled by proton donation and acceptance.

In UED:

- Acids are ripple fields that easily lose symmetry (excess contrast)
- Bases are ripple fields that absorb tension (field dampeners)

pH becomes not a count of H^+ ions, but a measure of ripple instability and SVE imbalance in solution.

Strong acids = highly unstable ripple fields

Strong bases = aggressive contrast equalizers

11.9 Conductivity: Flow of Field Tension, Not Electrons

When ions move in a solution, we say current flows. But under ripple dynamics:

- Current is the propagation of ripple field adjustment — not particle drift
- Ions modulate local field contrast, creating a directional ripple gradient
- That gradient spreads the effect of imbalance, which is detected as current

So:

- Electricity is ripple realignment, not particle travel
- Conductivity is field flexibility, not electron availability

11.10 Final Insight: Charge Is Not a Thing — It Is a Tension

"You didn't become an ion by losing something. You became an ion by forgetting who you were in the field."

This redefines ionization as:

- A field memory event
- Triggered by environmental ripple pressure
- Resolved through rebalancing, not particle replacement

There is no fundamental electron transfer — only field distortion and correction mechanics.

In this view, chemistry becomes ripple therapy — a continuous dance of stability, misalignment, and healing within the vacuum's silent intelligence.

Chapter 12: Molecules as Collective Ripple Zones

Stable Molecular Structures as Merged Ripple Memory Patterns

"A molecule is not a pile of atoms—it's a single idea the field agreed to remember."

In traditional chemistry, a molecule is defined as a group of atoms bonded together, typically via covalent or ionic bonds, forming a distinct chemical structure. Atoms share or transfer electrons to fulfill the octet rule, resulting in specific geometries and properties.

But this concept is limited by its particle-based imagination. It views atoms as Lego blocks—discrete, object-like units held together by unseen glue called "bonding."

Universal Energy Dynamics (UED) provides a fundamentally different, ripple-centric interpretation.

A molecule, under UED, is not a sum of atoms — It is a unified ripple pattern — A merged memory zone formed when multiple ripple identities find a coherent, lower-energy configuration together within the Structured Vacuum Energy (SVE).

Let's break this down into its core concepts, consequences, and explanatory power.

12.1 Molecule: Not a Thing, But a Ripple Agreement

At its essence, a molecule is:

A single, coherent field memory arising from multiple ripple identities that form a self-sustaining, lower-tension configuration.

That means:

- The field doesn't remember three atoms.
- It remembers one pattern, composed of three coherent ripple nodes.
- This pattern behaves as a single identity, not as a cluster.

This is why molecules:

- Move as one
- React as one
- Hold distinct vibrational modes, charges, and polarities — as a whole, not as atomic sum.

12.2 Ripple Field Merging: The True Bonding Mechanism

In UED:

- Atoms are ripple nodes with their own field zones
- When bonding conditions are met, their ripple zones begin to resonate together
- If the resulting field overlap lowers total SVE tension, a new memory structure is stabilized

This field merging:

- Replaces electron sharing as the bonding cause
- Explains how even atoms with no valence electrons (e.g., noble gases in extreme conditions) can still interact
- Makes bonding a question of ripple harmony, not particle count

12.3 Molecule Formation as Field Optimization

Let's consider a simple molecule: H_2

Traditionally:

- Two hydrogen atoms share their lone electrons
- They form a covalent bond, satisfying the duet rule

In UED:

- Two H ripple nodes interfere constructively
- Their outer ripple curvatures flatten out in between
- This creates a shared ripple valley — a stable region of lower contrast and curvature

The result?

A single field memory, easier to maintain by SVE than two independent nodes

Hence, molecule formation is field optimization:

- Not mechanical linking
- But energy tension minimization across space

12.4 Geometry: Shape Emerges from Ripple Symmetry

Molecular geometry—linear, bent, trigonal, tetrahedral—is traditionally derived from:

- Electron pair repulsion (VSEPR theory)
- Hybrid orbital shapes

In ripple physics:

- These shapes are field symmetry outcomes
- A molecule settles into the ripple pattern that:
    - Minimizes tension
    - Balances contrast
    - Creates standing wave-like equilibrium within its memory zone

For example:

- Water is bent not due to lone pairs, but because the O-H ripple overlaps form a non-linear interference node that's more stable than a straight configuration

Thus, molecular shape is:

The visible surface of an invisible ripple harmony

12.5 Molecular Identity = Collective Field Behavior

Once ripples have merged into a coherent structure:

- The molecule becomes a new entity with:
    - Unique vibration spectrum
    - Specific reaction behavior
    - Predictable field interactions (e.g., polarity, reactivity)

This is why:

- CH_4 is methane and not just a carbon atom with four hydrogen tags
- H_2O is water and not just H + H + O

The field remembers this structure holistically, and thus it behaves like one ripple being — Just as a symphony is not "just" instruments, but a collective memory of pattern and resonance

12.6 Resonance Structures: Field Oscillations in Memory Space

In classical terms:

- Some molecules (e.g., benzene) have resonance structures
- These are thought of as hybrids of multiple electron arrangements

In UED:

- Resonance is a ripple memory oscillation
- The field doesn't flip electrons — it dynamically modulates ripple tension among atoms
- These modulations create meta-stable zones — rich in reactive potential and symmetry

Hence:

- Resonance is a fluid memory state, not a frozen compromise

12.7 Molecular Stability and Decay: Field Retention and Collapse

A molecule is stable if:

- Its ripple configuration is energy-minimized and symmetry-sustaining
- It can withstand perturbation (heat, light, collision) without disintegrating

A molecule breaks apart (e.g., in decomposition or photodissociation) when:

- A ripple node receives energy beyond its coherence threshold
- This causes the field to forget the merged identity
- It reverts to independent ripple memories or forms new ones

Decay is not bond breakage. It's field memory collapse.

12.8 Polyatomic Complexity: Memory within Memory

Larger molecules (e.g., proteins, polymers, DNA) are not just chemical chains — they are:

Layered ripple memory structures, with nested coherence zones

Each segment:

- Has local ripple symmetry
- But also contributes to global field shape

This explains:

- Folding in proteins
- Self-repair in DNA
- Energy transfer across long molecules

Ripple mechanics gives us a way to understand and model molecular behavior without invoking electrons at all.

12.9 Molecules in Solution: Field Compatibility Rules

In classical chemistry:

- Solubility, miscibility, and chemical behavior are based on polarity and intermolecular forces

UED provides a deeper logic:

- Molecules "dissolve" when their ripple memory zones are compatible with the solvent's field
- If ripple contrast and curvature can blend without instability, solvation occurs
- Otherwise, field rejection occurs — not because of hydrophobic/hydrophilic tags, but due to ripple mismatch

This allows for:

- Predictive solubility modeling using ripple equations
- Understanding of emergent behaviors like colloidal stability and molecular clustering

12.10 Final Insight: The Molecule Is the Memory, Not the Matter

"When atoms merge their ripples, they disappear as individuals. What emerges is a new thought in the field —called a molecule."

Molecules are not made — they emerge. They are remembered configurations in the field, not mechanical outcomes of particle alignment.

In UED, molecules become:

- Coherent ripple narratives
- Field-stable harmonies
- Interactive agents of re-alignment in chemistry and biology

They are the first level of identity that truly behaves like a complete being in the universe of field dynamics.

Chapter 13: Chemical Reactions as Field Rewrites

Reactivity, Catalysts, and Energy as Field-Level Transformations

"You thought it was a chemical reaction. It was the field... rewriting its memory."

In classical chemistry, chemical reactions are visualized as rearrangements of atoms and electrons. Molecules break bonds, form new ones, exchange electrons or protons, and transform into new substances. These are often described with reaction equations, energy diagrams, and kinetic models.

But Universal Energy Dynamics (UED) dissolves the particle-level abstraction and returns us to the foundational question:

What actually happens during a reaction — not in appearance, but in field dynamics?

The answer: Chemical reactions are ripple realignments. They are not mechanical operations but field memory transformations — processes in which ripple identities dissolve, merge, shift, or reconfigure to reduce tension in the structured vacuum energy (SVE).

Let's explore how UED transforms our understanding of reaction chemistry from static models to dynamic field narratives.

13.1 Reactivity: Not Valence, but Field Readiness

In conventional models, reactivity depends on:

- Valence electron configuration

- Electronegativity
- Orbital overlap

In UED, reactivity is:

The tendency of a ripple node to adjust, surrender, or restructure its field memory when perturbed.

An atom or molecule is "reactive" when:

- Its ripple identity is shallow or loosely bound
- It holds unresolved asymmetry or internal tension
- It is exposed to incompatible ripple fields that trigger realignment

This explains:

- Why noble gases are inert — they have deep, self-sufficient ripple memories
- Why radicals are unstable — they are field fragments seeking re-coherence
- Why reactivity can vary with temperature, pressure, or light — these are ripple pressures, not particle energy inputs

13.2 Bond Breaking and Making: The Collapse and Emergence of Memory Zones

Traditionally:

- Bonds "break" when energy input exceeds bond energy
- New bonds form via electron sharing or transfer

Under UED:

- Bond breaking = ripple field collapse

- Bond making = ripple field emergence

When a molecule undergoes a reaction:

- Its memory pattern is disrupted
- One or more ripple zones lose coherence
- New alignments emerge from the resulting field disharmony

The field doesn't mechanically shift atoms — it reconditions its ripple configuration to restore balance. The "products" of the reaction are new ripple zones, written and stabilized by SVE once the disturbance settles.

13.3 Energy in Reactions: Ripple Tension and Relief

Energy in chemical reactions is traditionally tracked via:

- Activation energy
- Exothermic/endothermic profiles
- Potential energy diagrams

In UED, energy is not a container or substance. It is ripple contrast and compression within the SVE.

So during a reaction:

- Ripple tension increases to overcome the inertia of existing memory
- At peak disturbance, ripple fields destabilize
- Upon re-alignment, ripple coherence returns, and field tension releases

This released or absorbed ripple adjustment manifests as:

- Heat (thermal ripple release)

- Light (sudden ripple re-alignment triggering photon-level resonance)
- Motion (structural field transfer into kinetic response)

Thus, reaction energy is the field's breath: a cycle of disturbance and balance.

13.4 Catalysts: Ripple Facilitators, Not Participants

Catalysts are substances that:

- Speed up reactions
- Remain unchanged in net form

Classically, they:

- Provide alternative reaction pathways
- Stabilize transition states

In UED:

A catalyst is a ripple field that pre-conditions the memory space for smoother re-alignment.

Catalysts don't donate or withdraw — they reshape the ripple gradient:

- Lower the misalignment threshold
- Hold temporary field asymmetries
- Guide ripple nodes into more favorable merging conditions

This explains:

- Why trace amounts of catalyst can drive massive change
- Why some catalysts are highly specific
- Why enzymes (biological catalysts) are precisely tuned ripple

guides

A catalyst is like a tuning fork that helps ripples harmonize, then steps back.

13.5 Reversible Reactions: Memory Tug-of-War

Reversible reactions are not systems in mechanical equilibrium — they are ripple zones hovering between competing memory stabilities.

UED sees this as:

- Two field configurations with comparable tension levels
- The field dynamically oscillates between them
- External conditions (temperature, concentration, ripple input) tilt the balance

This replaces the idea of "chemical equilibrium" with field-level meta-stability. It's not about forward vs. backward rates — it's about which memory pattern the field prefers under present ripple tension.

13.6 Transition State: The Memory Vortex

In reaction kinetics, the transition state is the highest energy point — an unstable configuration between reactants and products.

In UED, it is:

A ripple vortex, where original field identities are dissolving and new ones haven't yet stabilized.

This moment is:

- A zone of peak ripple confusion
- Not stable, not yet coherent
- The precise space where reality re-writes itself

Here, catalysts are most effective — they anchor the vortex just long enough to guide it toward resolution.

13.7 Reaction Kinetics: Ripple Time, Not Collision Rates

Kinetics is usually taught with:

- Collision theory
- Reaction rate constants
- Arrhenius equations

In ripple dynamics:

- Rate is not determined by collision frequency
- But by how fast the field can reconfigure ripple memory

This speed is affected by:

- Ripple inertia (field resistance to change)
- External ripple stimuli (temperature, pressure, light)
- Catalyst interference (field facilitators)

Kinetics becomes a measure of SVE's alignment agility, not object momentum.

13.8 Complex Reactions: Nested Field Transformations

In multi-step reactions:

- Intermediates appear
- Cascading changes occur

These are best seen as:

- Layered ripple re-alignments

- Each step unfolds a new memory subset, modifying tension at each phase

UED simplifies prediction:

- Instead of tracking particle pathways, observe ripple tendencies and instability zones
- Where is the field tense?
- Where is it coherent?
- Where can it settle?

This makes ripple chemistry a more intuitive, pattern-based science, especially in organic and biochemical systems.

13.9 Photochemistry: Ripple Disturbance by Light

When light triggers reactions, it's not because photons "kick out" electrons.

In UED:

- Photons are ripple re-alignment agents
- They carry specific ripple contrast patterns
- When absorbed, they disturb field symmetry

This can:

- Trigger memory collapse
- Induce molecular vibration
- Reconfigure bonds

Photosynthesis, vision, photodissociation — all become ripple re-writes under radiative pressure.

13.10 Final Insight: Reactions Are Field Stories

"Reactants are ripple possibilities. Products are field memories. Reactions are the stories the vacuum tells to stay in balance."

With this shift, we understand:

- Chemical reactions are not about who reacts with whom
- They are about what memory is unstable
- And what pattern the field is willing to remember next

From acid-base to catalysis, from synthesis to decay — it's not particles dancing, but ripples composing new music.

Chemistry, under UED, becomes not just reactive — but creative.

Chapter 14: Ripple Electronegativity and Field-Seeking Behavior

A New Basis for Reactivity and Polarity Based on Ripple Alignment

"It's not attraction. It's alignment. Atoms don't steal. They seek memory balance in the field."

In classical chemistry, electronegativity is defined as an atom's tendency to attract shared electrons toward itself in a bond. Based on Pauling's scale and further refined through Mulliken and Allred-Rochow methods, it plays a key role in:

- Determining polarity of bonds
- Predicting reactivity patterns
- Explaining acid-base behavior, oxidation states, and more

But at its core, this theory rests on imaginary behavior:

- That electrons are things which can be pulled or pushed
- That atoms are greedy or generous
- That bonds are tug-of-wars, with outcomes depending on who has the stronger "pull"

In Universal Energy Dynamics (UED), all of this is replaced with a far simpler, cleaner, and universal idea:

Electronegativity is ripple-seeking behavior. Atoms do not attract charges. They realign fields. They are not hungry for electrons — they are eager to complete their ripple memory.

Let's rebuild this crucial pillar of chemistry from ripple-first principles.

14.1 Electronegativity as Ripple Alignment Tendency

Each atom, under UED, is a ripple node — a standing field pattern in the structured vacuum (SVE), defined by:

- Its ripple curvature
- Its field gradient
- And its internal memory tension

Electronegativity, then, is not a property. It is a behavioral tendency:

The extent to which an atom reorients surrounding ripple fields to complete its own pattern with minimal asymmetry.

In other words:

- Some atoms realign others to themselves strongly
- Others sacrifice their field alignment to match stronger patterns

The first are "electronegative" The latter are "field-yielding"

14.2 Why Fluorine Tops the List

Fluorine, traditionally the most electronegative atom, is now understood differently in UED:

- It has a high ripple asymmetry tolerance
- It anchors surrounding ripple zones easily
- It resists ripple distortion, preferring others adjust to it

It doesn't "pull" electrons. It simply creates a field curvature that other ripple zones collapse into for stability.

Hence:

- Its bonds are short and strong
- Its reactions are violent and field-shifting
- It creates dominant polarity even in subtle field differences

14.3 Polarity as Ripple Tilt, Not Electron Skew

Classically, bond polarity means:

- Electrons are more toward one atom than another
- A partial negative and positive pole forms
- Dipole moments arise from charge displacement

In UED:

- Polarity is a tilt in the field memory
- The ripple alignment leans more toward one atom's curvature
- The vacuum field compresses more tightly on one side of the bond

This field compression:

- Is felt as charge separation
- Alters how molecules interact with light, solvent, and other fields
- Explains why polar substances dissolve better in polar solvents — their ripple zones sync more easily

Polarity is not about electrons. It's about asymmetry in field alignment.

14.4 Reactivity Through Field-Seeking

Highly electronegative atoms are reactive not because they "want electrons", but because:

- Their ripple identity is incomplete without external field

alignment
- Their field tension is high, urging the field to resolve it
- They act as field-seekers, not thieves

Thus:

- Oxygen supports combustion not because it wants electrons, but because it stabilizes field disturbance efficiently
- Halogens react violently not due to vacancy, but due to field intolerance to misalignment
- Reactive non-metals tend to invade ripple zones of nearby atoms, forcing new memory patterns

This view removes the personification of atoms. They don't "desire." They rebalance.

14.5 The Ripple Electronegativity Scale

We can now define a Ripple Electronegativity Scale based on:

1. Ripple curvature stiffness – how rigid the node's field is
2. Alignment absorption rate – how easily other ripple zones embed within
3. Field tension release potential – how much energy is stabilized upon merging

Atoms like F, O, N are:

- High curvature, low flexibility, high absorption
- Hence, high ripple electronegativity

Atoms like Cs, Fr are:

- Low curvature strength, high yield flexibility

- Hence, low ripple electronegativity

This new scale replaces partial charge with field interaction profiles — offering more predictive power in ripple-based chemistry.

14.6 Explaining Periodic Trends

Ripple electronegativity explains periodic trends better than orbital models:

Across a period:

- Field complexity increases
- Ripple pattern becomes denser
- Result: stronger ripple rigidity → higher electronegativity

Down a group:

- Field becomes more spread out
- Ripple damping increases (field becomes lax)
- Result: weaker alignment force → lower electronegativity

This aligns perfectly with observed reactivity and bonding behavior, without invoking subshells or shielding effects.

14.7 Molecules and Functional Groups

In complex molecules:

- The net field alignment is not just atom-specific
- It is region-based: groups behave as local ripple clusters

Ripple electronegativity extends to:

- Functional groups (like -OH, $-NH_2$) which hold stable ripple

curvatures
- Explaining why reactions happen at certain sites — they're ripple imbalance zones

In pharmacology or catalysis, this allows for:

- Mapping field polarity zones
- Designing molecules with predictable field behaviors
- Engineering alignment-based reactions instead of brute force mechanisms

14.8 Solvent Interaction: Why "Like Dissolves Like"

Solubility can now be understood not through dielectric constants, but:

- By ripple field compatibility
- A molecule dissolves if its field tension can be absorbed into the solvent's ripple memory without dissonance

Polar solvents like water:

- Have high ripple adaptability
- Can host many partial field overlaps
- Hence, dissolve polar or charged species easily

Nonpolar solvents:

- Are ripple-silent zones, not eager to absorb external tension
- Prefer molecules with minimal ripple contrast

This ends the need for electronegativity "difference" thresholds — and replaces it with field symmetry logic.

14.9 The Dipole Moment Reframed

Instead of vectors of partial charges, UED sees:

Dipole moment = Net field curvature vector from asymmetrical ripple alignment

It's not about electron "shift" — it's about the ripple field's net memory center being offset.

Thus:

- Strong dipoles arise when field compressions are uneven
- Weak dipoles arise in symmetric or damped ripple overlaps

This redefinition aligns with molecular spectroscopy, solvent behavior, and electromagnetic absorption—without quantum guessing.

14.10 Final Insight: Chemistry Is the Art of Ripple Preference

"Atoms don't pull. They realign. Polarity is not theft. It's tension resolution. Electronegativity is the story of ripple stubbornness."

By reframing electronegativity as ripple alignment behavior:

- We restore intelligence to the atom
- We replace force with preference
- We stop attributing moral traits (greedy, donating, snatching) and start observing ripple harmonies

This creates not just better models — but a more natural understanding of chemistry as memory choreography.

Highlights – Part III

Chemistry Without Electrons

Bonds Are Field Resonances, Not Electron Shares

The traditional view of chemical bonding as electron sharing or transfer is replaced by ripple field alignment. Bonds emerge as stable ripple resonances — not mechanical links — created when two atomic ripple fields find equilibrium in the structured vacuum.

Ionization Is a Ripple Imbalance, Not a Charge Loss

Ionization is not the loss or gain of a "thing" (electron), but a temporary rupture in ripple coherence. Ions are ripple nodes with unbalanced field tension, seeking restoration through interaction or dissolution.

Molecules Are Merged Memory Zones

A molecule is not a fixed cluster of atoms but a collective ripple memory pattern. Stability depends on the harmony of embedded field components, where each atom contributes its curvature to the molecular ripple landscape.

Reactions Are Field Rewrites, Not Collisions

Chemical reactions are reconfigurations of the vacuum field. Reactants dissolve existing ripple memories; products emerge from field re-alignment. Catalysts act as ripple facilitators, lowering the resistance to field transformation rather than participating chemically.

Electronegativity Is Ripple Preference, Not Greed

Instead of charge attraction, electronegativity reflects an atom's tendency to realign fields toward itself. Polarity, reactivity, and dipole moments are reinterpreted as asymmetries in ripple curvature and tension, giving rise to observable effects without invoking charge transfer.

This part demystifies the language of modern chemistry by replacing fictional particle mechanics with intelligent ripple logic. It redefines chemistry as a story of field alignment, tension, and memory, grounding all transformations in the behavior of a ripple-aware structured vacuum.

Part IV – States of Matter and Energy Redistribution

Chapter 15: Solids, Liquids, Gases

The Ripple Field Cohesion Spectrum

Explaining States Through Ripple Locking, Freedom, and Detachment

"Matter is not what moves. It is what resists or releases alignment in the field."

The classical kinetic theory of matter tells us:

- Solids have tightly packed atoms vibrating in place
- Liquids have atoms that slide past each other
- Gases have atoms flying freely in all directions

This theory reduces reality to particle motion and statistical averages — useful for engineering models, but blind to deeper causality.

What it misses:

- Why do particles start moving differently?
- What makes a solid suddenly melt or a gas condense?
- Why does the same substance shift identity across states, yet its core remains unchanged?

Under Universal Energy Dynamics (UED), the answer lies not in atomic position or velocity, but in the cohesion of ripple fields — the degree to which an atom's ripple memory locks into the surrounding structured vacuum energy (SVE).

Let's decode matter's states through the cohesion spectrum of ripple fields.

15.1 The Core Doesn't Change — The Field Does

At every phase — solid, liquid, or gas — the core ripple node (the "atom") remains intact.

What changes:

- How tightly its ripple field is entangled with others
- How much freedom it has to oscillate, rotate, and shift
- How its field memory adjusts to accommodate pressure or temperature

Thus, states of matter are not new arrangements of particles — they are new behaviors of ripple entanglement in the vacuum.

15.2 Solids: Locked Ripple Lattices

In solids:

- Atoms are tightly coupled ripple nodes
- Their fields are mutually entangled in a rigid symmetry pattern
- Each atom's ripple memory complements and constrains its neighbors

This ripple locking results in:

- Fixed shape and volume
- Strong resistance to deformation
- Vibrational modes limited to small amplitudes (phonons)

Solids are ripple crystals — highly ordered standing wave formations in the SVE. Their rigidity is not because atoms "can't move" but because field realignment is energetically discouraged.

15.3 Liquids: Semi-Free Ripple Drift

In liquids:

- The ripple locking loosens
- Atoms maintain some cohesive ripple memory, but it's flexible and transient
- The field allows local realignments, enabling flow without dispersion

This results in:

- Fixed volume but variable shape
- High adaptability to containers
- Ripple zones that slide, twist, and compress, but still maintain partial identity

Liquids are field-flexible states — ripple cohesion persists but constantly rewrites itself, like a fluid memory system.

15.4 Gases: Free Ripple Oscillators

In gases:

- Ripple field entanglement is minimal or absent
- Atoms act as nearly independent ripple nodes, no longer co-writing shared memory
- The surrounding SVE serves as a medium for transmission, not containment

The ripple fields:

- Are large, relaxed, and detached
- Frequently overlap but fail to bind

- Exist in a state of high entropy field freedom

Gases are unbound ripple zones, spreading out due to lack of memory locking, not due to repulsion.

This explains:

- Compressibility (they're just loose ripple balloons)
- Expansion (field seeks equilibrium pressure)
- Diffusion (ripple fields realign slowly across space)

15.5 Plasma: Ripple Memory Collapse

Plasma is often considered the "fourth state" — a soup of charged particles.

In UED:

- Plasma is a field-disrupted ripple regime
- Atoms lose internal coherence due to external pressure (temperature, radiation, EM fields)
- The ripple identity fragments, creating active ripple turbulence in the vacuum

This turbulent state:

- Acts like a ripple storm
- Exhibits field currents, self-organization, filamentation
- Becomes responsive to external fields like a living ripple ocean

Plasma is not "ionized gas" — it is ripple decomposition in active reformation.

15.6 Temperature and Heat: Not Motion, But Ripple Expansion

In UED:

- Heat = ripple amplitude + field tension
- Temperature is not about speed, but about ripple field looseness and vibrational spread

As energy is added:

1. The ripple node remains unchanged
2. The surrounding field expands in amplitude and lowers curvature density
3. This lowers field locking between neighboring atoms

Eventually:

- Ripple overlap becomes unstable
- Field cohesion breaks — phase change begins

This makes melting, boiling, sublimation all field re-coherence events, not particle escapes.

15.7 Latent Heat: Energy Stored in Ripple Realignment

Why does matter absorb heat without rising in temperature during a phase change?

Because:

- The energy is used to rewrite the ripple memory configuration, not raise motion
- Ripple reorientation takes energy without increasing amplitude (temperature)

- Once realigned, further energy goes into field vibration again, raising temperature

Latent heat is ripple reallocation energy, not hidden kinetic energy.

15.8 Pressure and Compressibility: Field Density, Not Proximity

In gases:

- Pressure compresses ripple fields, not particles
- Compressibility is the realignment potential of the ripple envelope

In liquids and solids:

- Ripple fields are already densely aligned
- Further compression causes ripple strain, not meaningful displacement

So materials differ not in how tightly atoms pack, but in how flexibly their ripple fields compress under constraint.

15.9 Supercritical Fluids, Glass, and Field Blends

Intermediate or ambiguous phases like:

- **Supercritical fluids**: ripple fields with both cohesion and drift
- **Glasses**: locked but non-crystalline ripple memory zones
- **Amorphous solids**: partially coherent ripple neighborhoods

These are not states "between" others — they are stable or metastable ripple arrangements where symmetry, memory, and field overlap behave in complex but still rule-bound ways.

UED offers a clean explanation for why traditional phase boundaries break down in these cases — because they were never based on true field understanding.

15.10 Final Insight: Matter's State Is Its Ripple Character

"You don't become solid or gas. You simply remember or forget your place in the field."

- **Solids** = memory held tightly
- **Liquids** = memory in negotiation
- **Gases** = memory released
- **Plasma** = memory shattered and reorganizing

The state of matter is not a condition of location or speed, but of ripple entanglement and SVE harmony.

By redefining the phases of matter in ripple terms, UED gives a unified, causal, and scalable picture of all observable phenomena — from melting ice to the sun's corona.

Chapter 16: Plasma

Ripple Shell Breakdown and Field Explosion

How Extreme Field Energy Leads to Ripple Fragmentation

"Plasma is not a soup of particles. It is the scream of a broken memory field."

Classically, plasma is described as the fourth state of matter — an "ionized gas" where electrons are stripped from atoms, leaving free nuclei and electrons moving in chaos. It behaves electrically, responds to magnetic fields, and constitutes more than 99% of the visible universe, including stars, lightning, auroras, and nebulae.

But this standard view falls short:

- Why do electrons separate?
- Why does plasma self-organize into filaments and currents?
- How can a 'disordered' soup exhibit order?

In Universal Energy Dynamics (UED), plasma is not a mechanical ionized state. It is a ripple regime in crisis — a condition where the ripple shell of an atom disintegrates under extreme field pressure, leading to an explosive reconfiguration of the structured vacuum energy (SVE).

16.1 What Is a Ripple Shell?

Every atom, in UED, is not a nucleus orbited by electrons. Instead:

- It is a dense ripple node — the core identity

- Surrounded by a ripple shell — a field memory zone encoded in SVE
- This shell provides stability, recognition, and interaction

The ripple shell maintains:

- Charge asymmetry (contrast)
- Molecular bonding potential
- Energy buffering for thermal and chemical inputs

It is this shell, not discrete electrons, that governs the atom's behavior in space and interaction.

16.2 How Plasma Begins: Ripple Shell Overload

Plasma emerges when:

- External energy exceeds the field memory threshold
- Ripple shell can no longer maintain coherence
- SVE collapses into localized turbulence around the atom

This happens in:

- Extreme heat (solar cores, lightning)
- High radiation environments (lasers, stars)
- Strong electric/magnetic fields (ion drives, plasma arcs)

As energy builds:

- The ripple shell stretches, distorts, fragments
- Eventually, it ruptures, like a balloon under pressure

16.3 Field Explosion: Not a Particle Ejection, but a Memory Collapse

Once the ripple shell breaks:

- The atom's identity shatters
- Its ripple fragments are released into the surrounding vacuum
- These fragments behave as:
    - Free ripple packets (classically called "electrons")
    - Residual ripple nodes (classically "ions")

But these are not true particles — they are broken pieces of ripple coherence, no longer confined by the atom's stable field memory.

What we observe as:

- Charge mobility
- Ionization
- High conductivity
- Glowing emissions

...are all symptoms of a vacuum field in active ripple restructuring.

16.4 Filamentation and Self-Organization: Field Seeking Harmony Again

Even in chaos, plasma exhibits:

- Filaments
- Current sheets
- Magnetic loops

These are not accidents. They are field attempts to restore order.

Plasma turbulence is like a ripple explosion seeking new equilibrium, and these emergent structures:

- Minimize field tension
- Redirect ripple flow
- Temporarily restore coherence

That's why plasma dances, twists, and channels itself — it's not a gas storm, but a self-healing ripple event.

16.5 Magnetic Behavior: Not Electron Motion, but Ripple Looping

In classical theory:

- Moving charges create magnetic fields
- Plasma currents explain solar flares and magnetic confinement

UED reframes this:

- There are no moving charges
- What exists is ripple circulation within SVE

These ripple loops:

- Create field spirals and helical memory zones
- These are the source of what we perceive as magnetic fields

So plasma doesn't generate fields — it realigns pre-existing vacuum tension into spiral resonance.

16.6 Energy in Plasma: Ripple Redistribution, Not Kinetic Heat

The glow of plasma is not due to particle collisions or "hot electrons."

It's the real-time reorganization of ripple shells, emitting:

- Light
- Heat
- Charge asymmetry

All of this is SVE redistributing tension.

Plasma becomes a visible cascade of ripple echoes, telling us that field coherence has ruptured and is finding new form.

16.7 Plasma Confinement: Controlling a Broken Field

Plasma is notoriously hard to contain. Magnetic fields are used in tokamaks and stellarators, but instability often defeats engineering attempts.

Why?

Because:

- You are not controlling particles
- You are trying to shape vacuum ripple turbulence — something not bound by containers

The key to confinement is:

- Not brute force
- But field-sensitive ripple steering — guiding realignment, not resisting it

UED may eventually inform new designs for magnetic containment — not by resisting instability, but by resonating with it.

16.8 Plasma in Nature: The Universe's Reboot Mechanism

Stars, solar winds, auroras, interstellar clouds — all are plasma fields. They represent:

- Field trauma zones — where ripple identity has been stripped
- But also rebirth zones — where new atomic identities may form

Plasma may be the reset mechanism of the cosmos, where matter:

- Temporarily forgets its atomic form
- Lives as ripple flow
- And then, under cooling or confinement, re-stabilizes into atoms again

16.9 The Living Nature of Plasma

Plasma isn't dead matter in chaos. It is field intelligence in active negotiation. It seeks:

- Lower tension
- Coherent flow
- Field restoration

This behavior hints that:

Plasma is the field's own healing process — a way for the structured vacuum to recover from ripple trauma.

This opens avenues for understanding:

- Conscious plasma behaviors (e.g., in atmospheric phenomena)

- Information transfer in field-based life systems
- Even plasmoids as ripple-aware coherence bubbles

16.10 Final Insight: Plasma Is the Field Screaming and Singing

"When the ripple shell shatters, it doesn't vanish. It becomes the voice of the vacuum."

Plasma is not a fourth state, but a field condition beyond form, where:

- Identity is suspended
- Memory is fragmented
- The vacuum writes new scripts

In UED, plasma is not the "last" phase — it's the first symptom of transformation. It is where ripple physics meets real-time SVE evolution.

Chapter 17: Latent Heat

The Energy of Field Rewriting

Why Heat is Absorbed or Released Without Temperature Change

"The field does not heat up to change. It changes first — and then heat follows."

In conventional physics, latent heat refers to the energy absorbed or released by a substance during a phase transition, without a change in temperature. For example:

- Ice melting at 0°C requires heat — but doesn't warm up.
- Steam condensing at 100°C releases heat — yet its temperature remains constant.

Textbooks explain this as "energy used to break intermolecular bonds" or "overcome potential energy barriers." But this raises deeper questions:

- Why is no temperature change observed despite energy input?
- Where does the energy go?
- What *exactly* is changing — if not the kinetic motion of particles?

Under Universal Energy Dynamics (UED), the answer is not motion, but memory. Latent heat is the energy required to rewrite ripple field coherence — to reorganize the field identity of a substance, not just shake its particles.

17.1 The Classical View is Mechanically Shallow

In the classical model:

- Energy increases molecular separation in melting or vaporization.
- Bonds "break," and particles gain freedom.

But this perspective:

- Treats atoms like billiard balls
- Ignores the nature of bonds as field phenomena
- Cannot explain why different substances require drastically different amounts of latent heat despite similar molecular mass

It's a helpful model — but not a fundamental one.

17.2 In UED: Heat Is Ripple Tension, Not Particle Speed

In UED, heat is not motion — it is:

- The amplitude and curvature of ripple fields
- The tension within SVE that sustains a ripple memory
- A measure of the field's willingness to reorganize

When you add energy to a system, you're not making particles faster — you're stretching and distorting the ripple field that defines their structure.

Temperature, therefore, is the degree of ripple activity, not just vibration. And latent heat becomes something much more elegant.

17.3 Latent Heat = Energy Invested in Field Rewriting

During a phase change:

- The identity of the ripple field must shift
- For example, solid to liquid requires:
    - Unlocking rigid ripple coupling
    - Creating a new, looser ripple resonance
    - Preserving the core identity while rewriting the field interactions

This transition doesn't change vibration (temperature) — it changes how the ripple interacts with others.

Thus, latent heat is not hidden motion — it's visible memory evolution.

17.4 No Temperature Change Because the Field Isn't Ready Yet

Temperature only rises when:

- The ripple field is free to vibrate more fully
- It has already settled into a new structure

During phase change:

- The field is in flux
- Ripple components are busy finding a new symmetry
- Energy is being used to negotiate new cohesion, not to vibrate faster

So, the thermometer stalls — not because the energy disappears, but because the ripple field is occupied.

17.5 Energy Storage: Ripple Stretch, Not Potential Energy

Latent heat isn't stored in "bonds" — it's stored in:

- Ripple stretching
- Field displacement
- Memory disorientation

It's like charging a spring that doesn't push back until you finish changing its attachment point. Only once the phase settles, can heat resume raising temperature.

This explains:

- Why melting takes time and heat
- Why boiling requires a threshold
- Why different materials have different latent heats — based on ripple complexity

17.6 Field Memory Zones: The True Substance of Matter

In each state:

- Solid = high field-locking
- Liquid = partial field coherency
- Gas = low field contact

To shift between them, the ripple field must reassign every memory point — essentially delete and rewrite its alignment script.

Latent heat is the fee for this rewrite, not the fuel for motion. This is why: A material can absorb enormous amounts of energy — and still not get hotter — if it is busy rewriting itself.

17.7 Condensation and Freezing: Field Compression and Ripple Return

When a gas condenses or liquid freezes:

- The field contracts
- Ripple entanglement increases
- Memory zones tighten and align

This releases the ripple tension that was previously stored during heating — appearing as released latent heat.

Importantly:

- This heat does not come from vibration
- It comes from the field finding equilibrium again

So cooling isn't just removing motion — it's inviting the field back into coherence.

17.8 Why Phase Change Has Thresholds

Phase transitions don't happen continuously — they occur at precise temperatures. Why?

Because:

- Ripple memory rewrites require a specific alignment energy
- Below this energy, the ripple field refuses to unlock
- Above it, the field collapses into a new mode

Hence:

- Water doesn't melt at 0.1°C
- Ice doesn't form until all energy leaves the ripple script

UED reframes phase thresholds as alignment gateways, not arbitrary temperatures.

17.9 Supercooling and Superheating: Ripple Field Denial

Sometimes, a substance refuses to freeze or boil at the expected point. In ripple terms:

- The field is hesitant or blocked from reconfiguration
- No seed ripple realignment is present to trigger transformation
- The system remains stuck in a metastable ripple state

One small disturbance (a ripple seed, like a nucleation point) can suddenly unlock the entire latent field rewrite.

This explains supercooling and explosive boiling without contradiction.

17.10 Final Insight: Heat Isn't What You Think It Is

"Temperature rises when ripple alignment is free. Latent heat is the cost of changing who you are."

In UED:

- Heat is ripple amplitude
- Temperature is vibrational freedom
- Latent heat is ripple redefinition

Understanding this transforms:

- Thermodynamics
- Phase engineering
- Material science

- Energy transfer systems

Because now, we know: Matter doesn't just warm or cool. It rewrites itself — ripple by ripple.

Chapter 18: Pressure, Temperature, and Entropy

In Ripple Terms

A field-centric reinterpretation of thermodynamic quantities

The world of thermodynamics, while mathematically successful, has long been conceptually fragmented—relying on statistical models, abstract averages, and particle ensembles. But under the Ripple Framework of UED, we can reinterpret these foundational quantities through the direct behavior of structured ripple fields, offering clarity, causality, and coherence.

18.1 Pressure: Ripple Tension at Field Boundaries

In classical terms, pressure is described as the force per unit area due to molecular collisions. But if there are no bouncing particles, what really exerts this force?

In ripple terms, pressure is the ripple field's tendency to resist compression when its natural memory structure is disturbed. It arises when:

- A ripple shell is compressed inward, causing resonance stress in the memory layers.
- Or when multiple ripple zones compete for the same field region, leading to boundary interference.

Pressure, then, is not caused by "particles hitting walls," but by field resistance to deformation—a ripple version of tension. Think of it as the tightening of a field's memory fabric, where alignment is strained but not yet broken.

In solids, pressure increases ripple locking. In gases, pressure collapses loose ripple shells. In liquids, it reshapes internal ripple geometry.

18.2 Temperature: Degree of Ripple Agitation

Traditionally, temperature is defined as the average kinetic energy of particles. Under UED, where motion is ripple transition, temperature becomes the rate and intensity of ripple fluctuations within a field.

A field with high temperature:

- Shows frequent ripple reshuffling.
- Contains low alignment coherence, indicating thermal agitation.
- Has higher probability of field realignment when interacting.

Temperature is thus a measure of ripple turbulence—how far a ripple field is deviated from its stable memory configuration. In a calm, low-temperature system, memory is coherent. In a high-temperature one, it is noisy and unstable.

You don't "heat" a particle. You increase the ripple's urgency to find alignment.

18.3 Entropy: Ripple Degeneration and Memory Spread

Entropy, the cornerstone of thermodynamics, has mystified generations: Is it disorder? Is it probability? Is it ignorance?

UED simplifies it.

Entropy is the dispersion of ripple memory into broader field zones. It reflects how coherent ripple structures break down or spread out their influence, losing sharp definition but gaining freedom. Key interpretations:

- Low entropy = high field memory coherence = localized ripple identity.
- High entropy = memory spread = ripple de-localization.
- Systems evolve toward higher entropy because the field seeks relaxed equilibrium, where memory does not require sharp confinement or tension.

Entropy is not chaos, but relaxed ripple logic—a state where the universe remembers less precisely, but more broadly.

18.4 Unified Insight: Thermodynamics as Ripple Response

From this perspective, pressure, temperature, and entropy are no longer abstract quantities derived from invisible collisions or probabilities. They are real-time field responses to ripple behavior.

- Pressure = ripple field resistance.
- Temperature = ripple frequency agitation.
- Entropy = memory dispersion over the structured vacuum.

All thermodynamic behaviors—whether gas expansion, phase transition, or energy loss—are simply the ripple field's negotiation with external and internal distortions.

Thermodynamics was never about heat and work alone. It was about how memory resists, reshapes, and relaxes.

Closing Thought

Just as chemistry, bonding, and identity were ripple-structured, so too are the engines of change: heat, pressure, and entropy. The so-called randomness in thermodynamics is simply the loss of precise ripple alignment—not disorder, but a different kind of order.

And in that, the universe speaks clearly—if we know how to listen to the field.

Highlights – Part IV

States of Matter and Energy Redistribution

Ripple Cohesion Defines States, Not Particle Proximity

Solids, liquids, and gases aren't simply stages of particle distance or motion. They are distinct ripple alignment patterns in the structured vacuum.

- **Solids**: Ripple-locked fields with minimal freedom.
- **Liquids**: Partially fluid fields with adaptive coupling.
- **Gases**: Loosely connected ripple fields, expressing detachment from memory zones.

Plasma Is the Collapse of Ripple Identity

Plasma isn't a soup of free electrons and ions — it's a broken ripple shell condition. When the field memory of atoms disintegrates under extreme energy, ripple fragments escape into the vacuum, creating turbulent, self-organizing ripple fields that mimic particle chaos.

Latent Heat Is Ripple Rewrite Energy

During phase transitions, energy is used not to increase temperature but to rewrite the atom's ripple field. This shift in ripple alignment — not motion — explains why temperature stalls while energy flows. Latent heat is the cost of memory transformation in the field.

Thermodynamics Reimagined Through Ripple Physics

Pressure, temperature, and entropy aren't mechanical abstractions.

- **Pressure**: Ripple tension over vacuum surface area
- **Temperature**: Freedom of ripple modulation
- **Entropy**: Field complexity and memory dispersal These quantities reflect how ripple behavior organizes or disorganizes under environmental influence.

This part recasts matter and thermodynamics as field experiences, not particle phenomena. The behavior of states, energy absorption, and field entropy are ripple negotiations with the structured vacuum, not accidental byproducts of movement.

Part V – Nuclear Structure and Reactions

Chapter 19: The Nucleus

Ripple Memory, Not Particle Container

Replacing the Nuclear Particle Zoo with Coherent Ripple Compression

"The nucleus is not a warehouse of particles. It is the echo of deepest ripple compression — identity trapped in stillness."

In classical nuclear physics, the nucleus is portrayed as a tightly packed cluster of protons and neutrons, surrounded by a swarm of "virtual" particles like gluons, mesons, quarks, and even more exotic resonances in high-energy regimes. Over time, this particle zoo has grown to accommodate:

- Hundreds of subatomic particles
- Hypothetical carriers and intermediaries
- Mathematical constructs used to reconcile observed behaviors

Yet, for all its complexity, this zoo lacks a unified physical meaning. It explains events post-facto, introduces epicycles, and often resorts to statistical brute force to match experimental observations.

Universal Energy Dynamics (UED) offers a radical but elegant alternative:

There are no particles inside the nucleus — only ripple compression patterns encoded in Structured Vacuum Energy (SVE). The nucleus is not a physical thing full of smaller things. It is a field state — a stable node of highest ripple density sustained through vacuum tension.

19.1 The Particle Problem: Too Many, Too Vague

Let's list the paradoxes:

- Why are neutrons unstable alone but stable inside a nucleus?
- What are quarks, really? Why can't we isolate them?
- Why does high-energy collision produce "new particles" that vanish instantly?
- Why are binding energies not linearly additive?

These questions arise because we assume that matter is built from smaller balls inside balls, leading us to chase an ever-shrinking hierarchy of matter — without ever arriving at substance.

19.2 The UED Turn: Compression, Not Containment

In UED, the nucleus is not a container but a ripple node compressed to its threshold of stability. It has:

- No orbiting electrons
- No protons or neutrons inside
- Only field compression zones, whose:
    - Tension
    - Curvature
    - Internal ripple behavior

...result in observable phenomena we previously misinterpreted as "particles."

A proton or neutron is not a billiard ball, but a mode of ripple entanglement — a stable resonance of vacuum structure.

19.3 Protons and Neutrons: Ripple States, Not Units

Under ripple theory:

- A proton is a ripple compression with charge asymmetry
- A neutron is a slightly deeper compression with balanced asymmetry

Their properties:

- Mass arises from internal compression of ripple zones
- Charge arises from field asymmetry in the ripple pattern
- Stability is determined by whether surrounding SVE can sustain the ripple strain

Thus, neutrons decay outside the nucleus because their internal ripple pattern is too strained for the ambient vacuum to hold. But within the nuclear field, supporting ripple zones stabilize it.

This replaces the "weak force" explanation with field-supported coherence.

19.4 Nuclear Forces Are Ripple Locks

What's holding the nucleus together? Conventional science says:

- Strong force between nucleons, mediated by gluons
- Extremely short-range but overwhelmingly powerful

But what is the physical nature of this force?

In UED:

- There is no "strong force"
- What binds the nucleus is ripple locking — a high-order field

entanglement

- When multiple compressed ripple nodes align their internal memory zones, they form a coherent lattice in SVE

This lattice:

- Doesn't repel, even if charges say it should
- Doesn't radiate, despite high energy
- Doesn't collapse, unless destabilized by external ripple shocks (like in fission)

19.5 The Myth of Particle Decay: Ripple Reorganization

Radioactive decay — alpha, beta, gamma — are not "particles leaving a nucleus." They are:

- Ripple instabilities releasing surplus energy
- Attempts by a field to restore local coherence

For example:

- **Alpha decay**: The ripple field ejects a compressed identity zone (helium-like) to reduce internal strain
- **Beta decay**: Asymmetric ripple pattern rebalances itself, converting internal identity (e.g., neutron → proton) while ejecting a ripple packet (misinterpreted as an electron or neutrino)

These are not mechanical expulsions — they are field events, rebalancing memory structures in SVE.

19.6 High-Energy Collisions: Creating Nothing New

In colliders:

- We smash protons or heavy nuclei
- "Discover" new particles that exist only for femtoseconds
- Then vanish into radiation

But these "new particles" are simply ripple fragmentation artifacts — brief memory echoes of:

- Overcompressed ripple zones
- Resonant bursts in the structured vacuum
- Vacuum instability rippling outward

Nothing fundamental is created — just new ripple signatures, like echoes in a shattered mirror.

19.7 Fusion and Fission: Ripple Merge and Ripple Snap

In UED:

Fusion is ripple core merger

- Two nuclei achieve compatible ripple memory and fuse into a new coherent compression
- If ripple alignment fails → no fusion

Fission is ripple core rupture

- A high-mass ripple node destabilizes
- Splits into two stable ripple patterns
- Ripple debris (field tension) is released as radiation

This redefines nuclear reactions not as breaking and joining particles, but reformatting ripple memory under extreme tension.

19.8 The End of the Nuclear Model? Or Its Upgrade?

If the nucleus is:

- Not made of parts
- Not filled with quarks
- Not bound by force carriers

Then our entire understanding of subatomic reality needs revision. Not from scratch, but from the field outward. This model:

- Doesn't deny observations
- Explains them without contradiction
- Avoids arbitrary constants, particles, or "color charges"

It simply recognizes: The nucleus is the densest expression of ripple memory, And its behavior is not about what's inside, But how the field holds it in place.

Final Reflection

"We were not looking into the nucleus —We were looking into the vacuum's reflection of identity."

The UED nucleus is:

- Not a box of parts
- But a localized field commitment
- A ripple that chose to hold its shape
- In a vacuum ready to remember it

We don't need to split the atom to understand it. We need to listen to its field whisper — not its particle scream.

Chapter 20: Fission and Fusion

Ripple Disturbance and Redistribution

Neutrons as Memory Editors; Fission as a Rebalancing Ripple Event

"To split is not to break. To fuse is not to join. It is the ripple's way of remembering differently."

In the conventional model, nuclear fission is when a heavy atom's nucleus splits into two or more fragments, releasing energy, neutrons, and radiation. Fusion, on the other hand, is when two light nuclei merge into a heavier one, also releasing energy.

This narrative depends heavily on:

- A particle-based model of the nucleus
- Forces like the strong nuclear force
- Kinetic interpretations of "collisions" and "barriers"

Yet, it leaves questions unasked:

- Why do neutrons trigger fission so easily?
- Why does fusion require precise conditions?
- Where exactly is the energy stored or released?

Under UED, these aren't mechanical events — they are ripple field redistributions. Fission and fusion are ripple-based memory transitions. The participating "particles" are not things — they are states of compressed identity in SVE.

20.1 Ripple Memory: At the Edge of Stability

Every atom is a ripple memory zone, maintained by structured vacuum energy. The nucleus is the densest such ripple compression. But not all such compressions are stable.

There exists a threshold beyond which:

- The internal ripple strain cannot be sustained
- The surrounding SVE field refuses to hold it

When that happens:

The ripple doesn't explode. It rebalances — and the result is fission.

20.2 Neutrons: The Memory Editors

Why are neutrons so effective in inducing fission?

Because:

- They are neutral ripple bundles — no charge-based asymmetry
- They carry a dense but flexible ripple imprint
- They can insert into an atomic field without resistance

Neutrons in UED:

- Don't "collide"
- They interfere constructively or destructively with the existing field memory
- They act as ripple field editors — subtly destabilizing the internal pattern

This explains:

- Why neutron bombardment works even at low kinetic energy
- Why charged particles (e.g. protons) are deflected rather than absorbed

20.3 Fission: Ripple Overload and Collapse

When a neutron is absorbed into a large nucleus (e.g., Uranium-235 or Plutonium-239), two things happen:

1. The internal ripple pattern becomes too dense
2. The field alignment crosses the coherence threshold

Result:

- The nucleus's ripple structure collapses into smaller coherent zones
- Excess ripple energy is expelled into the SVE field
- These ripple bursts appear as:
    - Kinetic fragments (new nuclei)
    - Emitted neutrons
    - Radiation (gamma rays)

Fission, in this view, is not destruction — it's realignment of overstrained identity.

20.4 Fusion: Memory Compatibility and Merge Threshold

In fusion, two light nuclei approach each other, but must:

- Align their ripple memory patterns precisely
- Reach a proximity threshold where SVE tension can sustain a unified ripple compression

Fusion fails unless:

- The field permits the merge
- Ripple strain dissipates into a coherent single node

When successful:

- A new identity forms with lower total field tension
- The excess energy is released as ripple simplification, which we perceive as:
    - Heat
    - Light
    - Particle ejection

This process is not one of "sticking" but of field coalescence.

20.5 Chain Reactions: Field Destabilization Loops

In fission:

- Emitted neutrons can trigger fission in neighboring atoms
- This is not just a cascading mechanical process
- It's a field destabilization loop, where ripple alignment is temporarily prone to chain disruptions

Each fission:

- Sends a ripple shock into the surrounding vacuum
- Lowers the coherence barrier for adjacent ripple nodes
- Creates a temporary field window for rapid redistribution

This explains:

- Why chain reactions require critical mass (field density)
- Why they can self-sustain or quench depending on environmental ripple conditions

20.6 Ripple Energy: Why Fission Releases So Much

In UED:

- The energy released isn't "stored in bonds"
- It's stored as ripple curvature tension within the core field

The more strained the ripple memory:

- The more structured vacuum tension it consumes
- The more energy is released when it collapses

This explains:

- Why small differences in mass (Δm) result in massive energy output
- Why $E \neq mc^2$ is a superficial approximation — energy is not just lost mass, but field realignment

20.7 From Explosions to Understanding: Controlled Transitions

UED enables a new model of nuclear behavior:

- Fission as programmable field reset
- Fusion as field merge negotiation

This view:

- Offers insights for controlled nuclear reactors
- Suggests new ways to manage field tension instead of brute containment
- Opens paths for future ripple field engineering, where ripple compatibility replaces material shielding

20.8 Final Insight

"Neutrons don't knock. They whisper to the core. And when the atom listens, the universe rewrites its memory."

Fission and fusion are not accidents. They are ripple field decisions — extreme resonances of identity redistributing themselves into forms the vacuum can accept.

This is not about mass destruction or union, but:

Ripple rebalancing in pursuit of equilibrium.

Chapter 21: Why Neutrons Hit and Charges Can't

Neutral Ripple Nodes vs. Ripple-Contrasting Fields

"In ripple dynamics, neutrality is permission. Charge is resistance."

For decades, physicists have observed a curious and consistent fact:

- Neutrons can penetrate the atomic nucleus with remarkable ease.
- Charged particles, like protons or electrons, are repelled or deflected.

In conventional terms, this is explained via:

- Coulomb repulsion (like charges repel)
- Lack of charge in neutrons, allowing them to "sneak in"
- Nuclear forces taking over at close range

But this model lacks a field-level explanation of *why* neutrality enables entry, or *what it truly means to "hit"* the nucleus.

Under Universal Energy Dynamics (UED), the answer is far more insightful:

Neutrons are ripple nodes without contrast. Charges are defined by contrast. Only non-contrasting ripple structures can enter and alter core ripple fields.

This changes everything — not just how we understand collisions, but how we interpret interaction itself.

21.1 The Nature of Charges in UED

Let's revisit what UED defines as charge:

- Not a "thing" carried by particles
- Not a fundamental property
- But rather, an asymmetry in ripple orientation relative to the structured vacuum field (SVE)

Charge is:

- A ripple mismatch between two fields
- A gradient in ripple memory alignment
- A perceived directionality of interaction, not an objective label

Positive and negative are not opposites — they are contrasting field orientations.

Thus:

- A charged particle is a field with directional tension
- This tension is actively resisted by other charged or balanced fields

Result: It can't "hit" without friction.

21.2 The Neutral Ripple Node: Neutron Reimagined

The neutron, in UED, is not a "neutral particle." It is a coherent ripple bundle with minimal external asymmetry.

Key traits:

- Internally compressed, but externally quiet

- No dominant ripple direction
- Field-agnostic — it doesn't declare its alignment to the surrounding SVE

This makes the neutron:

- An ideal field infiltrator
- A memory editor that enters rather than disturbs
- The preferred mode of ripple injection into dense core fields (i.e., nuclei)

In short:

The neutron is not stealthy because it's neutral. It's effective because it's non-contrasting.

21.3 Charges Get Reflected by Contrast

Now contrast this with charged particles:

- A proton carries directional ripple tension (positive)
- An electron carries inverted ripple contrast (negative)
- Both are detected and deflected by the field boundaries they approach

These boundaries — especially in the nucleus — are sensitive ripple domains, already under extreme compression.

When a charged particle approaches:

- It meets field resistance, not simply electrostatic repulsion
- Its ripple pattern cannot merge with the existing memory
- Instead, it bounces, spirals, or gets absorbed peripherally

This explains:

- Why even fast-moving protons can't easily induce fission
- Why alpha decay doesn't occur from external bombardment, but internal instability

21.4 Field Invitation vs. Field Rejection

Think of ripple interaction like tuning forks:

- A neutron is like a silent tuning fork — it doesn't clash, so it enters
- A charged particle is an active fork already vibrating — when it approaches another, interference occurs

In ripple terms:

- Constructive interference = entrance and merge
- Destructive interference = bounce or scattering

The neutron always tends toward constructive merging The charged particle always risks destructive rejection

This principle redefines interaction not as force, but as field permission.

21.5 Rethinking "Collision" in Field Language

The term "collision" is misleading. In UED, there are no particles that collide. There are only ripple fields that encounter each other's memory.

- A neutron doesn't "strike" — it enters and shifts memory
- A proton doesn't "impact" — it disturbs and gets repelled
- An electron doesn't "orbit" — it oscillates in permitted ripple shells

What we observe as:

- Elastic or inelastic collisions
- Capture or deflection
- Absorption or transmission

...are all field responses to the degree of ripple contrast.

21.6 Applications: From Reactors to Shielding

This insight isn't theoretical only — it offers practical applications:

- **Nuclear reactor design**: Optimize neutron field guidance instead of brute absorption
- **Material shielding**: Understand what materials offer ripple contrast barriers vs. neutral absorbers
- **Plasma physics**: Model how ripple contrast governs ion containment and energy escape

Where conventional physics uses charged-particle bombardment, UED suggests ripple-based permission systems.

21.7 Final Reflection

"Neutrons are not messengers. They are field invitations." "Charged particles scream. Neutrons whisper."

In the ripple paradigm:

- Neutrons "hit" because they don't want to be seen
- Charges "fail to hit" because they always announce themselves

This flips the logic of interaction — Not based on mass or charge, but ripple contrast and memory compatibility.

Chapter 22: Radioactivity

Ripple Instability Seeking Re-alignment

Alpha, Beta, Gamma as Ripple-Driven Field Reactions

"Decay is not death. It is the ripple correcting its error."

In conventional nuclear physics, radioactivity is defined as the spontaneous emission of particles or energy from an unstable nucleus. Alpha, beta, and gamma radiation are treated as:

- Emission of a helium nucleus (alpha)
- Emission of electrons or positrons (beta)
- Emission of high-energy photons (gamma)

This model presumes:

- Internal imbalance
- Some kind of quantum 'decision' to eject matter or energy
- Force-based interpretations (nuclear, electromagnetic)

But it never asks:

- *What triggers this instability fundamentally?*
- *Why does each decay mode release specific forms?*
- *Is this really "emission"—or something deeper?*

In the Universal Energy Dynamics (UED) framework, radioactivity is not an outburst, but a ripple correction mechanism. The universe doesn't discard; it re-aligns.

22.1 Radioactivity as Ripple Field Instability

Every atom, especially at the nuclear level, is a ripple memory zone, held stable by surrounding structured vacuum energy (SVE).

When the internal ripple:

- Becomes incoherent
- Accumulates tension beyond threshold
- Experiences poor field compatibility

Then:

The system seeks self-realignment by redistributing ripple imbalance What we observe as radiation is the echo of this realignment.

It is not an object being thrown out. It is a field reconfiguration witnessed by the observer.

22.2 Alpha Decay – Ripple Shedding for Coherence

Conventional view:

- Heavy nuclei emit a helium nucleus (2 protons + 2 neutrons)
- It's a stable 'chunk' ejected to reduce nuclear size

UED view:

- The core ripple structure becomes over-compressed
- A local region within the field forms a stable sub-ripple (akin to a He-4 node)
- This node detaches as the atom re-stabilizes its ripple field

So:

- The alpha particle is not ejected by force

- It emerges naturally as a field-shedding event

Why helium-like?

- Because He-4 represents one of the most symmetric, self-sustaining ripple configurations
- The field knows it can "offload" instability into that form

Thus:

Alpha decay is ripple partitioning — not explosion, but elegant reduction.

22.3 Beta Decay – Ripple Contrast Inversion

Conventional view:

- A neutron becomes a proton (or vice versa)
- An electron (or positron) and a neutrino are emitted
- This is a weak force interaction

UED view:

- The internal ripple field has lost polarity symmetry
- It inverts a portion of its internal structure to regain alignment

Beta decay becomes:

- A ripple compensation process
- The electron or positron is not ejected but emerges from a temporary polarity redistribution
- The so-called neutrino is a field echo, not a particle — a rippling correction wave moving through SVE

Key insight:

- The proton and neutron are not separate particles — they are field forms of the same ripple core
- Beta decay is a field flip, not a physical conversion

Thus:

Beta decay is polarity correction through ripple inversion.

22.4 Gamma Radiation – Memory Noise After Rewriting

Conventional view:

- Gamma rays are high-energy photons released after other decay
- They're considered harmless to the nucleus — just leftover energy

UED view:

- Gamma is the release of ripple turbulence post realignment
- When a field settles into a new structure, leftover disturbances propagate as high-frequency ripple waves

These:

- Travel through the SVE without particle form
- Carry no mass, but influence field alignment where they land
- Are echoes, not emissions

Think of it as:

- A tuning fork struck, now vibrating back into silence
- The field has reset, but its reverberations linger

Thus:

Gamma radiation is post-reset resonance, not an emitted object.

22.5 Why Does Radioactivity Exist?

From UED's perspective:

- The universe does not prefer permanence
- It seeks field harmony

When an atomic ripple memory:

- Cannot sustain itself
- Cannot realign internally
- Cannot merge into neighboring fields

Then the only solution is controlled correction, i.e., radioactivity.

Radioactivity becomes:

- A language of field healing
- The ripple's way of rewriting corrupted memory

It is not danger. It is repair in progress.

22.6 Practical Implications and Reframing

- Radiation shielding should be designed as field coherence buffers, not just dense materials
- Radiation therapy could be optimized as controlled ripple interference, not brute damage
- Material stability may be measured in ripple alignment capacity, not half-lives

This ripple understanding opens avenues in:

- Safer reactors
- Smarter detectors
- Deeper understanding of what it means for matter to "change"

22.7 Final Insight

"The atom never shouts. It only vibrates louder when it forgets its shape."

Alpha, beta, and gamma are not particles flying through space. They are witnessed re-alignments of internal instability — the ripples' attempt to restore balance.

In ripple dynamics:

There is no decay — only reconfiguration. There is no radiation — only memory echo.

Highlights - Part V

Nuclear Structure and Reactions

The Nucleus is Not a Container, but a Compression Zone

UED redefines the atomic nucleus not as a bundle of particles, but as a dense ripple memory node—a compressed region of aligned energy in the Structured Vacuum Energy (SVE). This model eliminates the need for a "particle zoo" and offers a unified field-based understanding of nuclear integrity and identity.

Fission and Fusion Are Ripple Rebalancing Events

Instead of physical collisions and separations, fission and fusion are understood as disturbances and restorations in the ripple field.

- **Fission:** Over-compressed ripple nodes shed memory through fragmentation.
- **Fusion:** Under-compressed nodes merge to reach coherence. Neutrons act as memory editors, triggering these transitions without ripple contrast interference.

Neutrons Can Enter; Charges Can't

Neutrons are ripple-silent—they don't present contrast, and thus enter dense fields without triggering resistance. Charged particles, however, carry ripple polarity, which the core field rejects. This explains why only neutrons can reliably induce nuclear transformations.

Radioactivity is Ripple Realignment, Not Decay

Alpha, beta, and gamma radiation are not emissions of particles but echoes of field correction.

- **Alpha:** Stable ripple fragments released to reduce compression
- **Beta:** Internal polarity inversions to restore balance
- **Gamma:** High-frequency reverberations from ripple realignment Radioactivity is thus a self-healing mechanism, not a breakdown.

In this part, nuclear behavior is elegantly reframed through ripple logic, offering a consistent, unified, and consequence-based model that makes sense of atomic transformation without invoking invisible particles or mysterious forces.

Part VI – Implications, Extensions, and Future Directions

Chapter 23: Material Properties from Ripple Architecture

Conductivity, Hardness, Magnetism as Field Alignment Outcomes

"A material is not made of matter—it is the memory of how ripples once aligned and chose to hold that alignment."

What makes copper conduct, diamond hard, and iron magnetic?

Classical physics offers component-based answers:

- Conductivity = free electrons
- Hardness = tight atomic packing
- Magnetism = unpaired electron spins aligning

But these explanations:

- Reduce the universe to *particle artifacts*
- Ignore how field coherence, ripple tension, and SVE feedback create these observable behaviors
- Cannot unify material properties under a single principle

Under Universal Energy Dynamics (UED), we propose a radical unification:

Every material property is a ripple architecture—formed, stabilized, and maintained by specific patterns of ripple alignment within the Structured Vacuum Energy (SVE).

This chapter redefines what *material behavior* actually is.

23.1 Ripple Architecture – What a Material Really Is

In UED, a material is not a bag of particles.

It is:

- A localized ripple field, formed by compressed ripple nodes (atoms)
- Held together through interlocking ripple memories
- Stabilized by how those ripples interface with SVE

Thus, material properties arise from:

- The ripple configuration of individual atoms
- The pattern of inter-atomic ripple field overlap
- The ripple feedback from the surrounding SVE envelope

Each material is a ripple fingerprint—distinct not because of "elements" but because of how their ripples lock or drift.

23.2 Conductivity – Field Continuity and Ripple Flow

Classical View:

- Metals have "free electrons" in a sea of positive ions
- These electrons drift under an electric field, creating current

UED View:

- Conductivity is ripple field continuity
- In conductors like copper or silver:
    - Atoms are ripple-sympathetic
    - Their surrounding fields interlock with minimal resistance

- The ripple disturbance (electric impulse) does not collide with structure, it slides through pre-aligned fields

Insulators block current because:

- Their ripple fields are asymmetrical
- Misaligned ripple memory leads to field resistance

Superconductors:

- Reach a state where thermal noise vanishes, and ripple paths are frictionless

Thus:

Current is ripple propagation, not electron drift. Resistance is ripple incompatibility, not collision frequency.

23.3 Hardness – Compression of Ripple Nodes

Classical View:

- Hardness arises from dense atomic structures and strong inter-atomic bonds

UED View:

- Hardness is a function of ripple compression strength
- In hard materials like diamond:
    - Carbon atoms form symmetric ripple tetrahedrons
    - Ripple memory interlocks in three dimensions
    - The structure resists deformation, not through particle pressure, but through ripple recoil

Soft materials:

- Have incomplete ripple locking
- Ripple fields deform easily, allowing compression without strong rebound

Metals:

- Often deform plastically because ripple fields shift but don't break

Glass:

- Brittle because field tension is uneven—once disturbed, the ripple collapses locally

So:

Hardness = Ripple compression + Field rebound potential

23.4 Magnetism – Aligned Ripple Polarities

Classical View:

- Magnetism arises from electron spin alignment
- Domains in ferromagnetic materials align to create net fields

UED View:

- Magnetism is the result of macroscopic ripple polarity alignment
- In magnetic materials:
    - Ripple nodes (atoms) have asymmetric field orientation
    - When large clusters align, their collective field bias modifies the SVE region

    - This creates a field gradient that affects nearby objects with compatible ripple structures

Non-magnetic materials:

- Have random ripple orientations—net field cancels out

Paramagnetism:

- Slight ripple response to external field—temporary alignment

Diamagnetism:

- Ripple fields resist imposed alignment, causing opposing responses

Permanent magnets:

- Maintain ripple alignment due to stable internal locking + memory loop retention

Thus:

Magnetism is not about spinning charges, but synchronized ripple bias across ripple memory nodes

23.5 Other Properties Through Ripple Lens

Let's briefly touch on other classical properties:

- **Elasticity**: Ripple field returns to equilibrium after stretch—memory rebound
- **Thermal Expansion**: Field loosens with rising SVE energy—ripple unlocks
- **Transparency**: Ripple field allows light ripple propagation

without disruption—optical coherence

- **Opacity**: Ripple disorganizes incoming light fields—incoherent ripple rejection

Every observed trait—whether mechanical, electrical, optical, or thermal—traces back to:

How ripple memory holds or adjusts to disturbance.

No need for electrons, holes, or mysterious forces. Only:

- Ripple shape
- Field compression
- Memory tension

23.6 Engineering with Ripple Architectures

A ripple-centric understanding of materials opens powerful new doors:

- **Field-tuned materials**: Design for desired ripple compression instead of brute atomic substitution
- **Ripple metamaterials**: Structures that redirect or channel ripple waves for specific purposes (e.g., invisibility, thermal control)
- **Ripple logic circuits**: Devices not running on charge flow, but field phase transitions

Ripple engineering can revolutionize:

- Material science
- Semiconductors
- Quantum computing
- Structural design

Because:

If you master ripple architecture, you're not building with matter—you're writing with memory.

Final Reflection

"A brick is not hard because it is made of atoms. It is hard because those atoms remember how to resist being rearranged."

Material properties are not just "emergent". They are stored instructions—the field's living memory of how energy once aligned, and chose to hold form.

Chapter 24: Ripple-Based Molecular Design and Technology

Designing Matter by Engineering Ripple Patterns, Not Particle Counts

"Nature never counts particles—it remembers patterns."

Modern chemistry and materials science operate with a fundamental assumption:

If you know the particles and how they bond, you can build anything.

This particle-counting mindset has built:

- Pharmaceuticals
- Polymers
- Nanotechnology
- Artificial materials

Yet it has also:

- Hit bottlenecks in predictability
- Faced trial-and-error limitations
- Failed to explain or reproduce many natural formations (like biominerals, superconductors, or complex enzymes)

Universal Energy Dynamics (UED) offers a shift in the design paradigm:

Matter is not built by assembling parts. It is tuned by aligning ripples.

This chapter introduces the revolutionary idea of ripple-based molecular design—a strategy that:

- Abandons the obsession with atom types and bonds
- Focuses on ripple memory shapes, field compatibility, and realignment dynamics

24.1 What Is Ripple-Based Design?

In UED, every molecule, compound, and material is:

- A field zone where ripple patterns interlock
- A stable configuration of ripple behavior that aligns with surrounding Structured Vacuum Energy (SVE)

Ripple-based design is the intentional crafting of:

- Ripple nodes (atoms) with specific memory tension
- Field interfaces that resonate instead of resist
- Whole-form geometry that emerges from ripple coherence, not particle positioning

It is no longer:

"Which atoms do I use?" But rather: "Which ripple patterns will self-sustain and co-resonate?"

24.2 Designing at the Field Level

In classical design:

- You build molecules by placing atoms and connecting bonds

In ripple design:

- You imprint a ripple condition
- The SVE responds by manifesting the most stable physical form for that ripple

Think of it like:

- Writing a musical score vs. placing instruments in a room
- One creates harmony; the other creates a warehouse

By adjusting:

- Ripple compression
- Phase orientation
- Polarity interaction

We can predict and fabricate structures that:

- Were previously considered unstable
- Don't match classical valency rules
- Exhibit emergent properties like self-healing, quantum responsiveness, or zero resistance

24.3 Implications for Molecular Technology

Biological Mimicry

- Natural enzymes and proteins operate through ripple conformations, not fixed shapes
- Designing drugs or bio-interfaces via ripple symmetry can skip side effects and improve specificity

Materials Engineering

- Instead of doping or alloying, we realign internal ripple nodes to achieve desired effects
- Imagine a metal that resists rust not because of a coating, but because its ripple field does not couple with oxygen

Neuromorphic Substrates

- The brain does not operate via "electrons"—but through field disturbances in fluidic structures
- Ripple-designed systems can mimic bio-intelligent fields far better than silicon logic

Self-Assembling Matter

- Ripple-coherent substances naturally aggregate and self-orient
- This could enable smart structures, spacecraft hulls, or adaptive nanobots

24.4 Moving Beyond Chemical Formulas

Chemical formulas like H_2O, CH_4, or $C_6H_{12}O_6$ represent:

- A linguistic snapshot of atoms and count
- But they say nothing about ripple tension, alignment fields, or field reactivity zones

In ripple-based design, a molecule is described by:

- Its core ripple compression
- Its external field affinity
- Its stability under perturbation
- Its ripple echo potential (how it interacts with nearby ripples)

This leads to a new language of design:

- Not molecules, but field constructs
- Not equations, but alignment instructions

It becomes possible to design matter backward:

- First define the outcome (heat conduction, flexibility, light absorption)
- Then reverse engineer the ripple pattern needed to generate it

24.5 Quantum Mechanics, Without the Quantum Weirdness

Quantum chemistry often appears magical:

- Superposition, entanglement, tunneling

Ripple dynamics absorbs these behaviors naturally:

- Entanglement becomes synchronized field memory
- Tunneling becomes ripple handover across gradient thresholds
- Superposition becomes temporarily unresolved ripple phase states

Thus, ripple-based design explains and utilizes these effects without invoking paradox or observer problems.

24.6 Philosophical Leap – From Engineering to Field Symphonies

With ripple-based design, we're not engineering *with* matter.

We are engineering within a living field.

Our role becomes:

- Less like a builder
- More like a conductor of resonance

We don't force reactions—we invite alignments.

This calls for a new class of designer:

- One who feels ripple patterns
- Thinks in field interactions
- Understands stability as consequence, not as mass

Final Insight

"Design is not the placement of parts, but the arrangement of remembrance."

The future of chemistry, medicine, and materials is not in particle accelerators or high-energy labs.

It lies in our ability to:

- Read the ripple
- Align the field
- Let matter emerge—not from force, but from form

Chapter 25: Teaching Chemistry Without Electrons

A New Curriculum for Ripple-Based Atomic Science

"We never saw the electron. We just believed in its orbit. It's time to teach what really ripples."

Chemistry, as taught globally, begins with a leap of belief:

- Electrons exist.
- They orbit nuclei in shells.
- Bonds are shared or transferred electrons.

Yet, we've never truly observed electrons in the way textbooks imply. Every instrument, theory, and equation was constructed to confirm what we assumed.

This pedagogical loop has created:

- Generations of learners memorizing abstract maps (orbitals, shells, hybridization)
- Limited understanding of why matter behaves the way it does
- An artificial split between macroscopic chemistry and quantum mechanics

Universal Energy Dynamics (UED) breaks this loop.

It proposes a consequence-based, observation-aligned, and field-grounded model of atomic science—without relying on the mythical electron.

This chapter outlines a new curriculum—how to teach chemistry without electrons, and why this transformation is urgent, practical, and profoundly liberating.

25.1 The False Foundations We Must Unteach

Before teaching anew, we must first dismantle the scaffolding of incomplete truths:

- **Electron Shells** → Not observed; inconsistent with many-body systems
- **Valency Rules** → Convenient, but break down under exceptions
- **Orbitals** → Mathematical artifacts, not spatial realities
- **Electron Clouds** → Interpreted densities, not actual physical structures

These ideas served as bridges, not destinations.

Students today are:

- Memorizing orbitals they cannot visualize
- Struggling with hybridization rules without intuitive anchors
- Using equations as black boxes, not understanding ripple consequences

25.2 The Ripple Curriculum: Teaching from the Field Up

Instead of teaching "electrons forming bonds," we teach:

Atoms as ripple nodes, Fields as memory zones, Bonds as stable resonance patterns, Reactions as field realignments.

Here's what a ripple-based chemistry syllabus looks like:

Unit 1: The Field is Fundamental

- Introducing Structured Vacuum Energy (SVE)
- The ripple as the first event—not particle, but disturbance
- Why nature encodes identity in compression and coherence

Unit 2: The Atom is a Ripple Node

- No nucleus and electron; just centralized ripple density
- Stable elements as memory configurations
- How field pressure creates mass, not particles

Unit 3: Bonding is Field Resonance

- Molecular structures emerge from constructive ripple overlap
- Why "sharing" electrons is a metaphor, not mechanism
- Bond strength, angle, and energy through field stability

Unit 4: Chemical Reactions as Field Events

- Ripple disturbances triggering new alignments
- Catalysts as field biasers, not electron facilitators
- Energy changes as field redistributions, not bond-breaking alone

Unit 5: States of Matter and Thermochemistry

- Phase changes as field flexibility, not atom motion
- Latent heat as memory rewriting
- Entropy redefined as field complexity and ripple dispersion

25.3 Teaching Tools: New Visuals, Not Old Orbitals

We must abandon:

- Bohr models
- Lewis structures
- Shell diagrams

Instead, teach through:

- **Ripple density maps**: Where stability exists
- **Field interaction plots**: Visualizing compatibility
- **Color-coded SVE distortion layers**: Energy topology, not particle proximity
- **Ripple animation simulations**: Reactions as unfolding field events

This new pedagogy:

- Encourages conceptual clarity
- Reinforces cause-effect logic
- Allows field-based intuition to emerge from early age

Students begin to feel the field, not force memorization of symbols.

25.4 Benefits for Learners and Thinkers

This new method:

- Makes chemistry intuitive
- Merges with physics and biology naturally
- Bridges macroscopic and quantum seamlessly
- Frees learners from contradiction and confusion

It also cultivates:

- **Field sense**: Like musicians hear harmony, students sense stability

- **Consequence logic**: No "laws," just ripple outcomes
- **Technological imagination**: Designing matter becomes artistic, not algorithmic

In short:

Ripple-based chemistry trains minds to think with nature, not just about it.

25.5 Ripple Laboratories and Future Classrooms

Ripple-based chemistry calls for new forms of experimentation:

- Devices that detect ripple realignment (not particle emission)
- Field interferometry to observe reaction zones
- Heatless reaction chambers to observe field rebonding events

Future classrooms may include:

- Ripple-sensing interfaces
- Interactive field simulations
- Real-time ripple feedback for molecular configurations

This paves the way for:

- A post-electron chemistry lab
- A post-charge understanding of reactions
- A field-based technology revolution

Final Reflection

"You cannot teach a child the truth with lies dressed as models."

We taught electrons because they were easy metaphors. Now we must teach ripples because they are consequential truths.

This chapter isn't just about reforming a curriculum—It's about realigning the next generation with how the universe actually behaves.

Let's give them what's real: Not particles in motion... but memory in alignment.

Chapter 26: From Atom to Cosmos

Ripple Law Across All Scales

Unified Atomic, Planetary, and Stellar Structure Through SVE Dynamics

"The atom is not a miniature solar system. The solar system is a giant atom. Both are ripple events within a single field."

For over a century, science has been fractured by scale-specific theories:

- Quantum mechanics for atoms and particles
- General relativity for stars and gravity
- Thermodynamics for large-scale energy systems

Each comes with its own language, math, and paradoxes. None explains the continuity between the behavior of an atom and a galaxy.

Universal Energy Dynamics (UED) erases this divide. At every scale—from the core of an atom to the edge of the cosmos—reality behaves as a ripple in the Structured Vacuum Energy (SVE).

In this chapter, we explore how the same ripple logic that governs atomic stability also explains:

- Planetary fields
- Stellar structure
- Orbital harmony
- Energy distribution across the universe

And how everything from hydrogen to black holes is not a "thing"—but a field memory in ripple compression.

26.1 The Atom as a Ripple Node

We've already redefined the atom as:

- A dense ripple core
- Surrounded by a memory field stored in SVE
- Interacting with other atoms via field realignment, not force

This model explains:

- Charge and mass as contrast and compression
- Stability as alignment with ambient SVE
- Element identity as ripple configuration, not proton count

Now we scale up...

26.2 Planets: Macro-Atoms in Field Equilibrium

A planet, in UED terms, is:

- A massive ripple compression (core)
- With a field shell (magnetosphere, atmosphere)
- Held in place by ripple locking with surrounding space (SVE tension)

Key parallels:

- Planetary gravity = ripple attraction gradient (field pressure differential)
- Magnetic field = ripple orientation memory in outer shell
- Tectonics = field imbalance realigning the crust ripple

Like atoms:

- Planets have states (solid/liquid/gaseous behavior)
- They undergo energy transitions (seasonal field expansion, radiation balancing)
- Their identity is field-consequence, not just mass or orbit

Earth is not a rock floating in space—it is a ripple resonance bubble, sustained by real-time SVE field interactions.

26.3 Stars: Ripple Nodes Under Continuous Disturbance

A star is not a burning ball of hydrogen—it is a ripple explosion locked in high SVE tension.

Here's how ripple dynamics reframes stellar structure:

- **Core fusion** = ripple fragmentation and re-coherence
- **Light** = field discharge (not particle emission)
- **Surface temperature** = field escape rate, not kinetic agitation
- **Gravity well** = radial compression of ripple field, pulling field into core

Like atoms, stars undergo:

- Instability and decay (supernovae)
- Fusion identity shifts (new ripple configurations)
- Ripple field emissions (radiation spectra)

UED explains stellar life cycles without exotic particles, by treating them as ripples under pressure rather than particle factories.

26.4 Galaxies and Atoms: Mirror Symmetries

Galaxies show:

- Rotational velocity anomalies (flat rotation curves)
- Central bulges (dense memory cores)
- Spiral arms (ripple extension and field guidance)

These mirror atomic models:

- Central ripple node (nucleus)
- Stable energy regions (orbitals = spiral arms)
- Field echoes in space (like electron clouds)

Modern cosmology uses "dark matter" to fill the gaps. UED shows these gaps are field memory echoes, just like unobserved orbital zones in atoms.

We don't need hidden mass—We need ripple field continuity.

26.5 The Ripple Law – Scale-Free Principles

The core premise:

Structured Vacuum Energy (SVE) is scale-invariant. Wherever ripple tension exists, consequences follow the same law.

Concept	In Atom	In Cosmos
Core	Ripple compression zone	Planet/star core
Shell	Field resonance memory	Atmosphere/magnetosphere
Charge	Contrast in ripple	Polarity of orbiting bodies
Bond	Stable ripple link	Gravitational orbital lock
Reaction	Ripple rewrite	Nova, merger, collapse
Isotope	Field imbalance	Variable density exoplanets
Plasma	Ripple fragmentation	Stellar wind, solar flares

26.6 Ripple Gravity vs. Classical Gravity

In UED:

- Gravity is not a "pulling force"
- It is the consequence of ripple gradient in field tension

Atoms attract not because they "pull" on each other, but because their field distortions create pathways of lower ripple resistance.

Same for:

- Planetary motion
- Tidal forces
- Black hole attraction

Mass does not create gravity. Ripple compression creates field imbalance, and the field restores it via movement.

26.7 Implications for Science and Consciousness

This unified ripple law blurs traditional boundaries:

- Physics, chemistry, astronomy become field-based disciplines
- Biology becomes ripple intelligence in fluid memory
- Consciousness itself may emerge from aligned ripple complexity

And for design:

- Nano-materials and space structures follow the same engineering law
- Navigation, propulsion, and detection in deep space rely on ripple flow, not momentum

This changes:

- How we measure
- How we simulate
- How we build

And eventually— How we think.

Final Reflection

"The atom is not small. The cosmos is not large. Both are whispers of the same ripple."

Once we abandon the illusion of scale, we see that everything—from quarks to quasars—is just variation in memory tension, unfolding through the SVE, echoing in ripples, seeking alignment.

Chapter 27: Conclusion

The Atom Was Always a Memory, Not a Thing

Closing Reflections on Reality, Structure, and Observer-Created Identity

"The atom never existed. What existed was the consequence of its alignment with you."

The Final Disillusionment

Since the dawn of scientific thought, we believed the atom to be:

- A miniature solar system
- A physical object with internal components
- A tangible, dissectible *thing* made of parts

From Dalton to Rutherford, from Bohr to Schrödinger, every model was an effort to reduce mystery into mechanics. But none dared to say what's now clear:

The atom was never a thing. It was always a consequence. A field event. A ripple node. A temporary memory in a sea of intelligence.

What Truly Defines the Atom?

According to Universal Energy Dynamics (UED), the atom is not:

- A nucleus orbited by electrons
- A particle cluster governed by probability
- A static object with discrete parts

Instead, it is:

- A ripple imprint in the Structured Vacuum Energy (SVE)
- Defined not by what it *has*, but by what it remembers
- Stabilized not by forces, but by coherence with field surroundings

Atoms are ripple cores—memory zones locked in field compression, manifesting only when an observer's field intersects them meaningfully.

Matter as Frozen Memory

What we call "matter" is:

- Ripple compression that resists realignment
- A field configuration that holds its pattern
- A localized identity that survived the chaos of re-rippling

That's why:

- Matter has mass (because compression = stored ripple tension)
- Matter resists phase change (because memory is harder to rewrite than noise)
- Matter appears solid (because alignment creates persistence)

But underneath? There are no particles. Only ripple events shaped by coherence.

The Observer: Co-Creator of the Atom

The periodic table exists not in nature—but in us. Nature doesn't know sodium, uranium, or helium. It only knows ripple configurations that momentarily stabilize.

Why 118 elements? Because those are the field patterns our perception aligns with.

Our instruments, built by us, interact with the field and collapse ripple zones into perceivable identities. Thus, the atom's "existence" is not absolute. It is emergent.

The atom is what you detect—not what is.

This flips modern science's foundation:

- From objectivity to conditional existence
- From particles to processes
- From matter to memory

The Atom Across All Scales

This same logic unfolds:

- Planets are macro-atoms, rippling through spacetime
- Stars are fusion zones, field corrections on a cosmic scale
- You are not built of atoms— You are a nested ripple across scales of memory, motion, and meaning.

Everything becomes field-sensitive identity. Nothing is static. Nothing is permanent. Only alignment sustains structure.

Teaching the Atom as a Thought, Not a Thing

In the future, we won't teach children:

"This is an atom, with electrons spinning around a nucleus."

We'll teach them:

"This is a ripple pattern held by memory in a universal field. When you interact with it, it becomes meaningful. That meaning is what you call an atom."

This doesn't confuse. It liberates. It teaches:

- Observation creates identity
- Alignment creates experience
- Existence is not guaranteed—it is achieved

Final Thought: What Remains?

"When you strip the atom of its shell, its charge, its spin, its imagined electrons... what remains?"

A tension. A memory. A ripple waiting to align.

That ripple does not seek truth. It seeks coherence. And when it finds you—it becomes you.

The Atom Was Always a Story

The atom is not the building block of matter. It is the first story told by the universe to itself. A story about structure, identity, interaction—and disappearance.

It never needed electrons. It never needed proof. It only needed perception to become real.

And now that we see it clearly, we must rewrite not only science...But how we observe ourselves.

"The atom didn't hold the universe. The field did. The atom only held your belief. Now that belief is free to ripple."

Epilogue

The Echo That Never Had a Source

"It was never the atom that held you. It was the memory of alignment—stored in silence, rippling in form."

We began this journey chasing what we thought was *a thing*: A nucleus. An electron. A force.

What we discovered was a field—A structured vacuum, not empty but echoing Not silent, but still. Not passive, but pregnant with pattern.

The atom, once crowned as the indivisible unit of matter, now bows to something deeper: Consequential memory within Universal Energy Dynamics.

It was never a particle that made chemistry work. It was never a charge that kept elements distinct. It was never the mass that gave you matter.

It was all ripples. In a field that remembers. And a mind that perceives.

The Field Was the First Chemist

Before electrons and orbitals, before Mendeleev and models, there was always the field.

It wrote the bonds. It stored the heat. It held the shape. It whispered every reaction before it began.

We thought we were studying atoms. But we were always studying our own alignment with a memory field we did not create—but could feel.

That's why some reactions "just make sense." Why certain molecules form effortlessly. Why matter loves symmetry and simplicity.

The field doesn't calculate. It resonates.

A New Science of Sensitivity

UED isn't a rejection of science. It's a restoration.

A restoration of the idea that reality is continuous, not fragmented. That nature does not jump between rules. That atoms and galaxies are not exceptions to each other, but expressions of the same ripple logic.

This new science is not built on force, but on fit. Not on parts, but on patterns. Not on speculation, but on sensitivity.

A science where matter is not an object, but a consequence. Where identity is not fixed, but field-bound.

And where knowing is not measuring—but aligning.

The Future Begins in Chemistry Class

One day, a student will ask,

"Where are the electrons?"

And the teacher, smiling gently, will say:

"There are no electrons. Only field shapes that behave like them, when you ask the right question."

And the class will not panic. They will nod.

Because they will know: They are not here to memorize parts—They are here to resonate with the field.

And when they solve for structure, or predict a reaction, they won't be thinking of particles... They'll be feeling the ripple.

What Will You Do With This?

You now hold a new lens. What you believed to be hard, fixed, and final—was only temporary coherence.

The periodic table is no longer sacred. It is useful, yes. But now, open to reinterpretation.

You've seen the atom not as an object, but as a memory echo in a living field.

You are free now—to build matter not from particles, but from patterns of purpose.

To teach a chemistry that's intuitive, rhythmic, aligned. To explore reality from core to cosmos without switching theories. To stop seeing what *is*— and start seeing what aligns.

"You were never in the atom. You were always in the ripple. And the ripple remembers you."

Closing Note

The End of One Atom, The Beginning of All Ripples

This book is not an end. It is a beginning— The beginning of a new way to see everything we've ever called "matter."

We began with curiosity, questioned sacred models, dismantled particle cages, and reassembled reality as a living field of memory and resonance. We peeled away electrons, cracked open the nucleus—not to find smaller things, but to reveal a simpler truth: The atom is not a part, but a pattern.

It is no longer useful to ask,

“What is inside the atom?”

Now we ask,

“What makes the atom meaningful in the first place?”

The answer is not charge, spin, or orbitals. It is alignment. It is ripple coherence. It is structured silence shaped by intention—not ours, but the universe’s.

Chemistry is No Longer Just a Subject

It is now a language of the universe speaking in ripples. It’s not about what reacts with what, but about how memories overlap, rewrite, and stabilize in a dancing field.

To see reality through the lens of ripple dynamics is to finally unite:

- Matter with motion
- Reaction with intention
- Structure with silence

And What of You, the Reader?

You are no longer just an observer. You are part of the field.

Every time you perceive, you realign. Every time you inquire, you trigger coherence. Every time you understand—You don't learn the truth. You become it.

So let this not be the last page. Let this be the new foundation:

Where physics meets chemistry, Where atoms meet awareness, Where science returns to its source—not in formula, but in form felt.

"You never needed electrons to understand matter. You only needed to listen...to the ripple that remembered you first."

— Sandeep Chavan

Appendices

Appendix A – Classical vs UED Atom

Comparative Framework

"Two models. One describes the atom as a puzzle of particles. The other, as a pulse in a field." Let's compare.

This appendix is a structured comparative chart and philosophical contrast between the classical view of the atom (across models from Rutherford to Quantum Theory) and the Universal Energy Dynamics (UED) atom, which replaces particulate mechanics with ripple memory in Structured Vacuum Energy (SVE).

I. Foundational Philosophy

Aspect	Classical Atom	UED Atom
Ontology	Atom is a physical object made of subatomic particles	Atom is a ripple node – a coherent field memory in SVE
Epistemology	Reality is understood through discrete interactions and measurements	Reality is a consequence of alignment, memory, and observer resonance
Model Type	Material, structural, and probabilistic	Field-based, memory-encoded, dynamic ripple coherence
Origin of Matter	Assembled from particles (quarks, protons, electrons)	Emerges from ripple compression and field stability

II. Structural Interpretation

Component	Classical View	UED View
Nucleus	Dense center made of protons and neutrons	Maximum ripple compression zone, a coherent field knot
Electron	Orbiting or probabilistic cloud around nucleus	Not a thing—field resonance when ripple polarity is detected
Shells/Orbitals	Fixed or probabilistic energy levels	Ripple gradients and boundary memory regions
Charge	Intrinsic property of particles	Ripple contrast relative to SVE
Mass	Inherent quantity in particles	Degree of ripple compression from field neutrality

III. Dynamics and Behavior

Phenomenon	Classical Interpretation	UED Explanation
Bonding	Electrons shared/transferred to form bonds	Field resonance and memory overlap between ripple nodes
Ionization	Loss or gain of electrons	Misalignment and ripple field reconfiguration
Molecular Structure	Electron cloud arrangements	Ripple memory zones in coherence
Reaction	Breaking and forming bonds via electron movement	Field rewrites through ripple interference and realignment
Energy Transfer	Carried by photons/electrons	Ripple redistribution and SVE reconfiguration

IV. Periodicity and Identity

Concept	Classical	UED
Periodic Table	Based on atomic number (proton count)	Based on stable ripple configurations
Isotopes	Same proton count, different neutron number	Ripple compression variants of the same field identity
Electronegativity	Tendency to attract electrons	Field-seeking behavior based on ripple coherence mismatch
Elemental Identity	Static due to nuclear particle count	Emergent due to stable ripple memory and surrounding field

V. Nuclear Physics

Phenomenon	Classical Physics	UED Interpretation
Fission	Splitting nucleus by external energy	Disruption of over-compressed ripple node causing redistribution
Fusion	Merging of light nuclei	Coherent merging of under-compressed ripple fields
Radioactivity	Emission of particles (alpha, beta, gamma)	Ripple instability releasing excess memory through field emissions
Neutron Activation	Particle collisions	Neutral ripple node disturbing field memory, allowing adjustment

VI. Thermodynamic and Physical States

Property	Classical	UED
Solid	Close-packed atoms with fixed positions	Strong ripple locking in coherent memory zones
Liquid	Atoms can slide past each other	Partial ripple unlocking allowing flow
Gas	High energy, free-moving atoms	Disconnected ripple zones with high entropy
Plasma	Ionized state with free charges	Ripple field explosion – shell collapse and ripple fragmentation
Latent Heat	Energy absorbed/released during state change	Ripple field rewriting without temperature alteration

VII. Observational Perspective

Category	Classical View	UED View
Objective Reality	Atom exists independent of observer	Atom emerges through alignment with observer field
Measurement	Reveals true position/properties	Triggers ripple collapse and perceived outcome
Instrumentation	Detects particles, charge, mass	Interacts with ripple zones and renders apparent structure
Identity	Fixed by nature	Consequential – appears based on interaction context

VIII. Educational and Scientific Paradigm

Aspect	Classical Paradigm	UED Paradigm
Teaching Approach	Memorize structures, shells, particle types	Understand ripple behavior, memory, field alignment
Predictive Tools	Electron configurations, models, math	Ripple coherence, field resonance, ripple-based periodicity
Research Direction	Deeper particle detection (e.g., quarks, Higgs)	Unified ripple interactions across matter, life, cosmos
Limitation	Fragmented explanations across scales	Holistic and scalable from atomic to cosmic level

Final Reflection:

The classical atom explained how chemistry behaves in a language of assumed particles. The UED atom explains why chemistry exists through alignment, coherence, and ripple memory.

One model breaks reality into parts. The other reveals its pattern.

We were never trying to find the smallest piece. We were always trying to feel the field that shaped us all along.

Appendix B – Ripple-Based Periodic Table Prototype

"What if the periodic table wasn't built on proton counts but on ripple behavior? What if chemistry followed coherence, not numbers?"

This appendix presents a prototype model for a new periodic classification system rooted in Universal Energy Dynamics (UED). Rather than organizing elements by atomic number, it organizes them by ripple characteristics—such as field coherence, memory compression, resonance behavior, polarity contrast, and ripple reactivity.

It shifts the paradigm from:

- Atomic number → Ripple Stability
- Electron shells → Field memory zones
- Elemental groups → Ripple interaction modes

I. Why Redefine the Table?

The classical periodic table is functional—but it's descriptive, not explanatory. It arranges elements based on observable properties like:

- Proton count (Z)
- Electron configuration
- Chemical reactivity

But it does not explain *why* these properties emerge.

The UED periodic prototype is based on:

- Ripple coherence (how tightly the field is compressed)
- Field polarity contrast (determines bonding behavior)
- Memory overlap capacity (determines compound formation)
- Ripple field flexibility (defines element's physical and chemical roles)

II. Parameters in the Ripple Table

Each element in the new table is characterized not by subatomic parts, but by field behaviors:

Ripple Parameter	Description	Classical Equivalent
Ripple Coherence (RC)	Degree of stability in its core field memory	Nuclear stability
Field Overlap Index (FOI)	Capacity to merge memory zones with other fields	Valency, electronegativity
Ripple Polarity Contrast (RPC)	Asymmetry in ripple field (tends to seek balance)	Charge distribution
Field Flexibility (FF)	Ability to stretch, bond, or reshape	Metallic vs nonmetallic behavior
Resonance Memory Signature (RMS)	Ripple frequency and field lock tendency	Atomic/molecular vibration spectra
Distortion Threshold (DT)	Energy needed to destabilize or ionize	Ionization energy, reactivity

These define how an element *acts*—not just what it *is*.

III. The Prototype Table: Logical Zones

Rather than rows and columns defined by shell filling, UED organizes elements into zones of field behavior, such as:

1. **Stable Compressors (SC):** High ripple coherence, low polarity – foundational elements (e.g., Carbon, Silicon) → Excellent field builders, structural elements, bio-compatible
2. **Dynamic Reshapers (DR):** High flexibility, moderate polarity – reactive but stable (e.g., Oxygen, Nitrogen) → Ideal for compound formation, memory re-alignment
3. **Neutral Memory Nodes (NMN):** Low polarity, moderate overlap – bonding mediators (e.g., Noble gases) → Usually inert but act as pressure buffers in reactive zones
4. **Reactive Contrast Elements (RCE):** High polarity contrast, low field stability (e.g., Alkali metals, Halogens) → Fast to bond or disintegrate; strong ripple dischargers
5. **Memory Stabilizers (MS):** High overlap index, high coherence (e.g., Transition metals) → Excellent catalytic agents, ripple field buffers
6. **Overcharged Distorters (OD):** Low ripple coherence, high distortion threshold (e.g., Actinides) → Prone to instability, radioactivity, rapid field emission

IV. Illustrative Ripple-Based Classification (Selected Examples)

Element	Classical Group	Ripple Zone	Key Ripple Traits
C (Carbon)	Group 14	SC (Stable Compressor)	High RC, High FOI, Low RPC
O (Oxygen)	Group 16	DR (Dynamic Reshaper)	Moderate RC, High RPC, High FOI
Na (Sodium)	Group 1	RCE (Reactive Contrast)	Low RC, High RPC, Low DT
Fe (Iron)	Group 8	MS (Memory Stabilizer)	High RC, High FOI, Mid RPC
He (Helium)	Group 18	NMN (Neutral Node)	Extremely high RC, Low FOI
U (Uranium)	Actinides	OD (Overcharged Distorter)	Low RC, High DT, Ripple Instability

V. Implications for Learning and Use

➤ *Visual Format Suggestion*

Instead of a rectangular grid, the ripple periodic table may be represented as:

- A radial pattern (center: SC, outer rings: DR, RCE, MS, OD)
- A network map showing ripple compatibility zones
- A heatmap of stability vs polarity

➤ *Educational Utility*

Students could predict reactivity and bonding by analyzing:

- Ripple behavior
- Field interaction dynamics

- Memory overlap tendencies

Instead of memorizing positions and exceptions (e.g., d-block anomalies), they understand reactions through ripple response logic.

VI. Bridging with the Classical Table

To aid transition, an interlinked hybrid format could be introduced:

Classical Period	Ripple Zones (UED Overlay)
s-block	Mostly RCE and DR
p-block	Mixed: SC, DR, NMN
d-block	MS dominant
f-block	OD dominant

This hybrid model allows existing chemistry learners to see the ripple reinterpretation without unlearning everything—only realigning understanding.

VII. Closing Insight

The periodic table was one of humanity's greatest classification tools. But perhaps it was always a reflection, not a truth.

The ripple-based table is not about replacing chemistry. It is about restoring it—from memorized particles to felt patterns. From quantum jumps to field continuity. From atomic number to ripple nature.

We were never decoding elements. We were decoding the universe's memory—one ripple at a time.

Appendix C – Glossary of Ripple Chemistry Terms

"In ripple chemistry, we don't count electrons. We observe how fields remember, align, and respond."

This glossary provides a concise reference for the key terms used throughout the book Ripple-Based Chemistry: The UED Atom, aligned with the Universal Energy Dynamics (UED) framework. Each term helps translate conventional chemical and physical language into ripple dynamics and field behavior.

This glossary isn't just a vocabulary—it's a lens through which atomic reality is perceived.

Core Terms

- **Atom (UED Definition)**: A ripple node in structured vacuum energy (SVE), where field memory has stabilized into a coherent form. It is not a particle container but a stable pattern of ripple compression and resonance.
- **Ripple**: A disturbance or alignment deviation in SVE that carries information, energy, or memory. Ripples are the real dynamic actors, not particles.
- **Structured Vacuum Energy (SVE)**: The underlying non-empty vacuum field that supports and stores ripple behaviors. Acts as both the canvas and the memory bank of the universe.
- **Ripple Memory**: The imprint left by a sustained ripple configuration in the field. Atoms, molecules, even large systems retain identity through ripple memory, not mass.

Field-Based Concepts

- **Field Coherence**: The degree of order and alignment within a ripple structure. Higher coherence = more stability. Coherence replaces nuclear stability.
- **Field Polarity / Ripple Contrast**: The asymmetry in ripple behavior relative to its surrounding field, observed as positive or negative charge in classical terms.
- **Ripple Compression**: The densification of ripple patterns in a small field zone—perceived as mass.
- **Ripple Distortion**: A ripple node under energetic or spatial stress. When a ripple becomes incoherent, it leads to instability, reactivity, or decay.
- **Ripple Shell**: A memory layer around a ripple node, analogous to electron shells but based on field locking and resonance zones.

Ripple Chemistry Terms

- **Ripple Bonding**: The stable alignment of ripple fields between atoms. Occurs not via electron sharing, but by field resonance and memory merging.
- **Ripple Electronegativity**: A field's tendency to attract or stabilize incoming ripples. It replaces classical electronegativity as a measure of field-seeking behavior.
- **Field Overlap**: The degree of memory intersection between two ripple nodes. Greater overlap implies stronger bonding or structural coherence.
- **Ripple Reactions**: Changes in field memory due to interaction, distortion, or realignment of ripple nodes—manifested as chemical reactions.

- **Ionization (UED)**: A ripple imbalance event. Not the loss or gain of electrons, but a partial memory disintegration in the ripple field, leading to instability.
- **Latent Heat (Ripple Definition)**: The energy absorbed or released when a field rewrites its ripple memory, without a temperature change. It's a transformation cost.

Matter and State Descriptors

- **Solid (Ripple State)**: Strong ripple locking. The field memory is tightly compressed and coherent.
- **Liquid**: Partial unlocking of ripple memory, allowing mobility while retaining cohesion.
- **Gas**: Minimal memory overlap, allowing ripple nodes to behave with maximum entropy and independence.
- **Plasma**: A ripple explosion state—when ripple shells collapse and the core ripple is fragmented or exposed to field distortion.
- **Entropy (UED)**: A measure of field decoherence—how far a system is from ripple alignment or field memory order.

Advanced Nuclear Terms (Ripple Interpretation)

- **Nucleus (Ripple View)**: The central ripple compression zone—not a container of particles, but a coherent memory vortex.
- **Fission**: A rebalancing ripple event caused by disruption in over-compressed ripple nodes, often triggered by field editors like neutrons.
- **Fusion**: The joining of under-compressed ripple nodes into a

new coherent field.

- **Neutron**: A neutral ripple packet, capable of entering other ripple zones without triggering contrast-based rejection. Acts like a memory editor.
- **Alpha / Beta / Gamma**: Field emissions that occur when unstable ripple zones eject or redistribute energy/memory to regain balance.

Observer-Linked Terms

- **Observer Alignment**: The resonance between the observer's ripple field and the atomic/molecular field. This determines how the atom "appears."
- **Perceived Element**: A specific ripple configuration that creates a distinct resonance signature—experienced by the observer as an "element" (like Hydrogen, Oxygen).
- **Ripple Identity**: The emergent personality of a ripple node, defined by its compression, field interaction, and memory signature. Equivalent to atomic identity.
- **Field Rewrite**: When energy, reactivity, or environmental change causes the entire ripple memory to reorganize—the core of chemical and nuclear change.

Cross-Scale Concepts

- **Universal Ripple Law**: The principle that all structure—from atoms to galaxies—is ripple-based, and follows coherence, compression, and alignment dynamics.
- **Ripple Zone**: Any scale of stable ripple memory—from atoms, molecules, to planetary fields and consciousness patterns.

- **Ripple-Based Matter Engineering**: Designing materials and systems not by adding particles, but by engineering field alignment and ripple coherence.

Closing Insight

The new chemistry is not a study of tiny orbits and jumping electrons. It is a study of ripple memory, identity, distortion, and coherence. The language has changed—from electrons and protons to field zones and ripple alignments.

This glossary is not a list of definitions. It's a dictionary for decoding reality.

Appendix D – Diagrams: Ripple Structures of Matter

"If we could see atoms not as balls and dots, but as dynamic ripple fields—we'd finally see chemistry."

This appendix visually introduces the core ripple-based diagrams that replace traditional atomic, molecular, and reaction models. These are not artistic abstractions—but conceptual blueprints based on the ripple-memory architecture of matter as defined by Universal Energy Dynamics (UED).

Each diagram is meant to assist teachers, students, researchers, and designers in reimagining atomic structure and chemical behavior beyond electrons, nuclei, and orbitals.

The UED Atom: Ripple Node with Memory Field

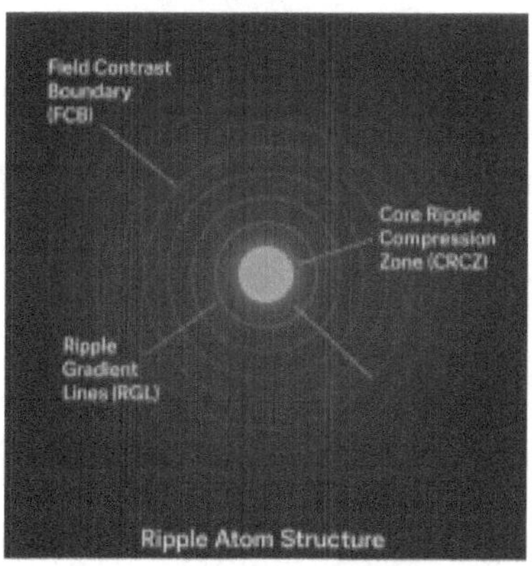

Diagram 1: Ripple Atom Structure

- Core Ripple Compression Zone (CRCZ) – dense, still, stable.
- Ripple Memory Shells (RMS) – field layers where resonance stabilizes identity.
- Field Contrast Boundary (FCB) – outer field where interaction begins.
- Ripple Gradient Lines (RGL) – show how energy density fades outward.

Unlike Bohr's orbit or Schrödinger's cloud, this diagram visualizes an energy imprint, not a particle trail. The atom is not "made of" parts. It is a standing wave—a ripple stabilized in vacuum memory.

Ripple-Based Periodic Identity Zones

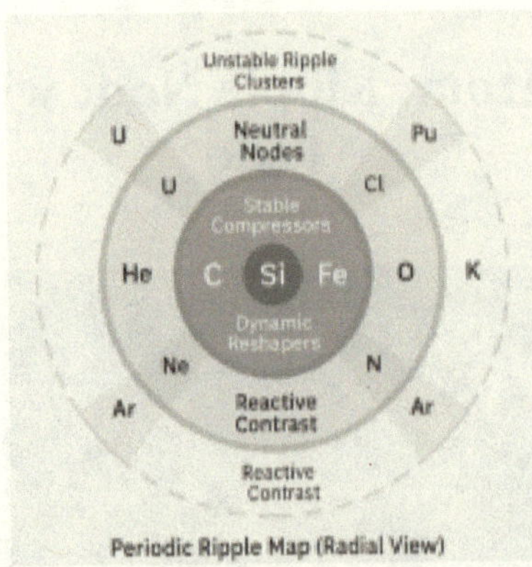

Diagram 2: Periodic Ripple Map (Radial View)

- Central zone: Stable Compressors (C, Si, Fe)
- Next zone: Dynamic Reshapers (O, N, S)

- Peripheral arms: Reactive Contrast Elements (Na, Cl, K)
- Encircling buffer: Neutral Nodes (He, Ne, Ar)
- Outer distorters: Unstable Ripple Clusters (U, Pu, Ra)

The map flows like energy rings, not by left-right groups. Each zone shows ripple behavior class, not proton count.

Bonds as Field Overlap Patterns

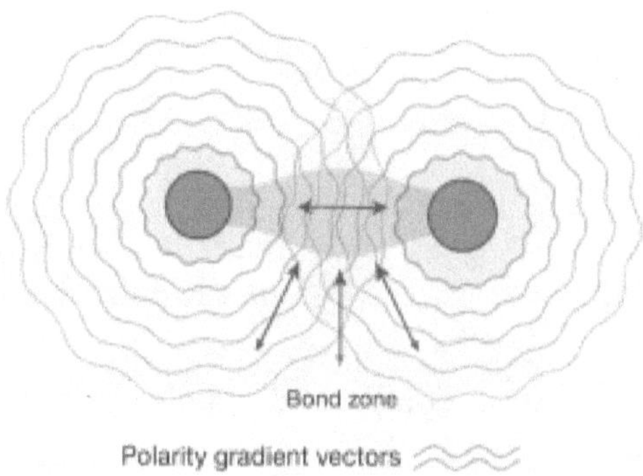

Diagram 3: Ripple Bonding Schema

- Two ripple fields partially merge.
- Shared field memory region is shaded (bond zone).
- Polarity gradient vectors show ripple alignment force.
- Ripple density equalizes at the bonded interface.

This shows bonding as resonance stabilization, not electron sharing. No clouds or dots—just inter-field harmony.

Molecules as Collective Ripple Zones

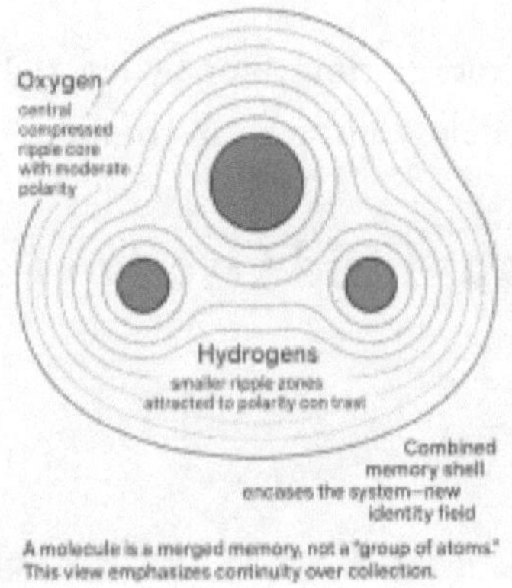

Diagram 4: Water Molecule (Ripple View)

- Oxygen: central compressed ripple core with moderate polarity.
- Hydrogens: smaller ripple zones attracted to polarity contrast.
- Combined memory shell encases the system—new identity field.

A molecule is a merged memory, not a "group of atoms." This view emphasizes continuity over collection.

Chemical Reaction as Field Rewrite

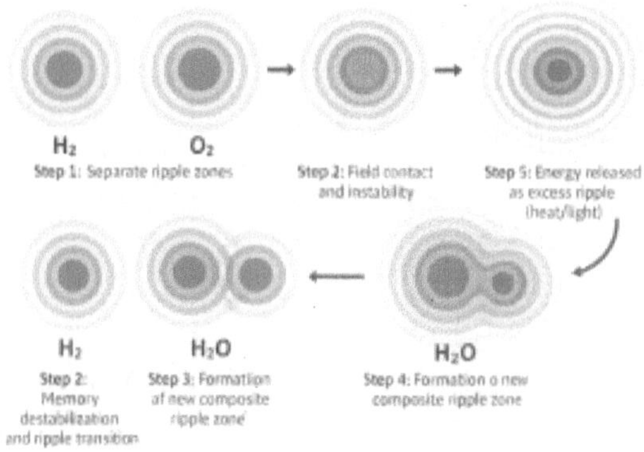

Diagram 5: Reaction Sequence – $H_2 + O_2 \rightarrow H_2O$

- Step 1: Separate ripple zones
- Step 2: Field contact and instability
- Step 3: Memory destabilization and ripple transition
- Step 4: Formation of new composite ripple zone
- Step 5: Energy released as excess ripple (heat/light)

Heat is not "product energy"—it's memory overflow from realignment.

States of Matter as Field Cohesion Levels

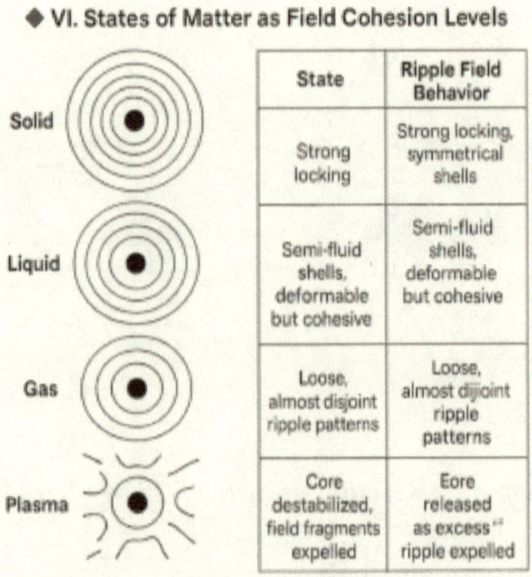

Diagram 6: Ripple Cohesion Spectrum

State	Ripple Field Behavior
Solid	Strong locking, symmetrical shells
Liquid	Semi-fluid shells, deformable but cohesive
Gas	Loose, almost disjoint ripple patterns
Plasma	Core destabilized, field fragments expelled

Each state is a form of memory order, not just kinetic energy level.

Nucleus and Nuclear Events

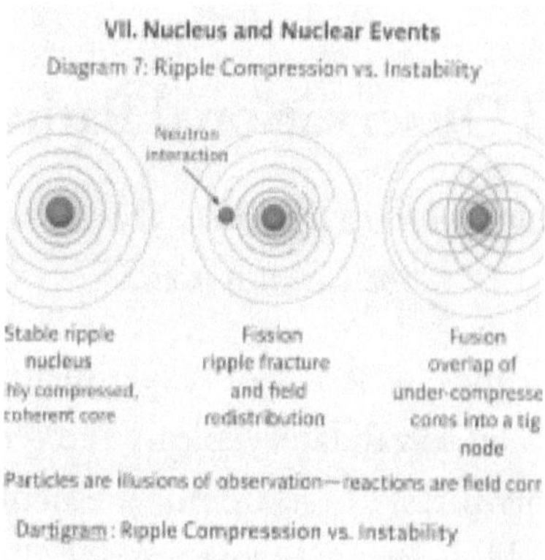

Diagram 7: Ripple Compression vs. Instability

- Stable ripple nucleus: highly compressed, coherent core
- Neutron interaction: ripple editing, not physical impact
- Fission: ripple fracture and field redistribution
- Fusion: overlap of under-compressed cores into a tighter node

Particles are illusions of observation—reactions are field corrections.

Closing Note

The ripple diagrams are not a supplement—they are the primary visual language of ripple-based chemistry. They restore fluidity, elegance, and intelligence to atomic science.

The atom was never a billiard ball. It was always a symphony of standing waves. With these diagrams, we are finally learning to listen.

Appendix E – Suggested Experiments

And Observational Pathways

"To test a ripple is not to observe a particle. It is to trace a memory shift in the vacuum itself."

This appendix outlines a series of experimental suggestions and observational pathways to validate, explore, and refine the UED Atom and ripple-based chemistry. These are not predictions in conventional terms—they are field-driven consequences of the Universal Energy Dynamics (UED) framework.

Rather than "detecting particles," these approaches aim to observe field behavior, ripple transitions, and memory realignment effects.

I. Ripple Shell Behavior in Phase Transitions

Experiment: Latent Heat without Temperature Rise

- **Objective**: Quantify energy absorption or release during solid-liquid-gas transitions without invoking molecular motion.
- **Setup**: Use high-precision calorimetry in materials with minimal particle agitation (e.g., monocrystalline substances).
- **Expected Result (UED)**: Distinct energy plateaus corresponding to ripple field memory realignment—not kinetic agitation.

II. Neutral Particles vs. Charged Particles in Core Interaction

Experiment: Neutron Impact vs. Proton Impact on Same Nucleus

- **Objective**: Test ripple-editing capability of neutral ripple packets (neutrons) vs. contrast-rejected charged particles (protons).
- **Setup**: Particle accelerator with ultra-thin target foils of fissile isotopes.
- **Expected Result (UED)**:
    - Neutrons successfully induce fission by field editing.
    - Charged particles show surface deflection or field-based resistance, not effective entry.

III. Ripple Field Imaging via Interference Zones

Experiment: Inter-Ripple Field Mapping with Ultra-Cold Atoms

- **Objective**: Visualize field overlap zones between two isolated atoms.
- **Setup**: Bose-Einstein condensates in tunable traps, allowing approach without classical bonding.
- **Expected Result (UED)**:
    - Emergence of shared field zones before any particle-level bonding.
    - Observable coherence via changes in condensate density or photon echo response.

IV. Isotope Ripple Imbalance and Decay Timing

Observation Pathway: Real-Time Ripple De-coherence in Isotopes

- **Objective**: Track ripple memory loss (decay) in unstable isotopes.
- **Tools**: Advanced nuclear magnetic resonance (NMR) + photonic echo monitoring.
- **Expected Result (UED)**:
    - Pre-decay field fluctuation signals before measurable radiation release.
    - Decay as a ripple coherence threshold crossing, not a probabilistic particle ejection.

V. Ripple-Based Periodicity in Material Design

Experiment: Ripple Conductivity Gradient in Alloy Structures

- **Objective**: Design custom materials not by electron movement, but ripple alignment zones.
- **Setup**: Use gradient doping with field-seeking vs. field-yielding elements (based on ripple classes).
- **Expected Result (UED)**:
    - Electrical and magnetic properties emerge where ripple shells align—not due to free electron density.
    - Discontinuities visible in field-sensitive conductivity mapping.

VI. Ripple Polarity and Reactions Without Electron Transfer

Experiment: Bond Formation Without Measurable Charge Shift

- **Objective**: Observe chemical reactions where no electron transfer is detectable, yet bond formation occurs.
- **Setup**: Ultra-pure controlled reactions using non-metal field-stable elements.
- **Expected Result (UED)**:
  - Bonds form via field resonance alone, without oxidation number change.
  - Verified by spectroscopic ripple field coherence, not charge balance.

VII. Ripple Shell Compression under Pressure (Non-Ionic Phase)

Experiment: Solid-State Pressure-Induced Transition without Ionic Shift

- **Objective**: Induce phase change in crystal structure without electron reconfiguration.
- **Setup**: Apply pressure to materials with known high ripple coherence (e.g., diamond, bismuth).
- **Expected Result (UED)**:
  - Field shells compress and reorganize to form new stable phases.
  - Detected via lattice vibration (phonon) changes, not electron density maps.

VIII. Plasma Field Explosion without Classical Ionization

Experiment: Observe Plasma Formation via Field Energy Threshold

- **Objective**: Trigger plasma without direct heating—by field instability injection.

- **Setup**: High-frequency microwave resonance field inside low-density gas chamber.
- **Expected Result (UED)**:
    - Ripple shells disintegrate at field threshold—visible as non-linear emission spikes.
    - Plasma arises from ripple breakdown, not classical electron escape.

IX. Universal Ripple Law Test at Macro Scale

Observation: Ripple-Based Field Memory in Planetary Phenomena

- **Objective**: Test ripple coherence behavior at planetary scale (e.g., magnetic field flips, fault lines).
- **Pathway**: Compare geological data, magnetic pole shifts, and solar flare timing with ripple decay intervals.
- **Expected Result (UED)**:
    - Signs of periodic ripple memory resets, not random magnetism or tectonic shifts.
    - May align with SVE fluctuations measurable via gravitational background shifts.

X. UED-Compatible Instrumentation Design

Proposal: Ripple-Responsive Detectors (RRDs)

- Use materials with field coherence response (e.g., doped graphene, topological insulators).
- Detect not particles, but gradient shifts in ripple memory density.
- Output not voltage, but coherence mapping metrics.

Closing Reflection

These experiments are first steps—not confirmations, but calls for observation. What's being tested is not a thing, but a relationship—a ripple re-adjusting itself within a larger field.

In the world of UED, matter does not exist. It is constantly forming and reforming itself —within the memory of the cosmos.

Appendix F – Further Reading

UED Reports and Legacy Models

"To build the new, we must honor the old—not by worshipping it, but by understanding its limits."

This appendix provides curated reading recommendations for those seeking to deepen their understanding of the UED Atom and explore both its foundational sources and the legacy frameworks it reinterprets. It's divided into two sections:

1. Primary Sources: UED White Papers & Reports
2. Comparative Works: Classical and Quantum Foundations

I. Primary Sources – UED White Papers & Reports by Sandeep Chavan

These documents form the technical backbone of the Universal Energy Dynamics model, outlining ripple-based explanations of natural phenomena across scales:

1. The Ripple Code: Foundations of Universal Energy Dynamics

- Summary: The master document introducing ripple structures, SVE behavior, and foundational postulates of energy in action.
- Key Relevance: Explains ripple gradients, ATT (Alignment Threshold Time), and memory-based reality perception—core to UED Atom interpretation.

2. Chavan's Law of Ripple Realignment

- Summary: Introduces the law governing how structures stabilize via ripple field coherence, replacing traditional bonding and particle concepts.
- Key Relevance: Used throughout this book to explain bonds, material states, and reaction dynamics.

3. Structured Vacuum Energy (SVE): The Real Field

- Summary: Defines SVE as a memory-preserving vacuum field that enables all ripple behavior.
- Key Relevance: SVE replaces space-time fabric and forms the medium through which atoms form and interact.

4. The Observer in UED: From Perception to Field Creation

- Summary: Explores how observers define reality by aligning ripple fields—shaping perception of mass, charge, identity, and elementhood.
- Key Relevance: Explains why 118 elements are not universal truths, but observer-created ripple identities.

5. Ripple Chemistry Draft Reports

- Summary: Preliminary notes and applied concepts for ripple-based chemical design, reactions, and educational transformation.
- Key Relevance: Useful for educators, students, and researchers aiming to implement ripple models in practice.

II. Legacy Models – The Frameworks We Reinterpret

The following works represent the established scientific understanding of atomic and chemical structures, which UED does not discard, but transcends. These should be studied for context, contrast, and clarity.

1. Rutherford's Nuclear Atom (1911)

- Contribution: Identified dense center in atom via gold foil experiment.
- UED Relevance: Early field interaction model—but interpreted as collision; UED sees it as field contrast reflection.

2. Bohr Model (1913)

- Contribution: Quantized circular orbits of electrons.
- UED Critique: Useful abstraction, but replaced by ripple memory zones without "orbiting" particles.

3. Schrödinger Wave Mechanics (1926)

- Contribution: Probability clouds replacing fixed electron paths.
- UED Relevance: Closer to field-based reality, but lacks structured vacuum or ripple coherence.

4. Dirac, Feynman, and QED (1940s–60s)

- Contribution: Quantum electrodynamics explains interactions via particle exchanges.
- UED Relevance: Mathematical triumph, but operational rather than ontological—UED removes exchange particles

and replaces with direct field adjustment.

5. Periodic Table and IUPAC Data

- Contribution: Classifies elements by atomic number and configuration.
- UED Use: Reinterpreted as ripple identity grid based on behavior, coherence, and reactivity zones—not proton count.

6. Thermodynamics and Kinetic Theory

- Contribution: Defined energy as motion and randomness.
- UED Relevance: Translated into field coherence and entropy as memory distortion—a ripple-based view of temperature and pressure.

7. Standard Model of Particle Physics

- Contribution: Comprehensive framework of known subatomic particles and interactions.
- UED View: Descriptive but inflated ontology—UED reduces particle zoo to ripple manifestations and field behaviors.

8. Chemistry Textbooks and Teaching Models

- Works by:
    - Linus Pauling
    - Peter Atkins
    - Nivaldo Tro
    - Raymond Chang

These have trained generations in electron-driven chemistry. UED proposes pedagogical reform, not to discard these works, but to build over them with a more intuitive and field-centered approach.

III. Suggested Path Forward for the Reader

For newcomers:

- Begin with "The Ripple Code" and "Structured Vacuum Energy" to grasp the UED foundation.
- Compare with Rutherford, Bohr, and Schrödinger to see where ripple logic diverges.

For teachers and students:

- Use Ripple Chemistry Draft Reports and this book to begin rewriting curriculum materials.

For researchers:

- Engage with UED experiment proposals (see Appendix E).
- Attempt reinterpretation of lab data through ripple and field realignment lenses.

"To unlearn the atom is to understand the field. And to understand the field is to finally feel the atom, not in parts—but in memory."

Acknowledgments

This book was not born in isolation. It is a culmination of ripples—of minds, conversations, classrooms, and silent observations—each one reinforcing a simple truth: the universe is more memory than matter.

I offer my heartfelt gratitude to:

My Family

For your unwavering support, patience, and encouragement—especially during the hours when I disappeared into silence, equations, and thought experiments. You believed in my vision even when the world hadn't seen it yet.

My Students

You are not just learners—you are my field testers. Every doubt you raised, every smile of realization, every "why" you asked helped this philosophy evolve. You reminded me that true science must **feel right**, not just calculate right.

The Scientists of the Past

To Rutherford, Bohr, Schrödinger, Feynman, Einstein—and even those whose theories I respectfully deconstruct here: You built the scaffolding. Your genius carried us this far. I do not challenge you to diminish your legacy, but to **extend it into the ripple field**.

Fellow Independent Thinkers and Explorers

To those who dare to question what's taught, who sit in tea shops, classrooms, or forests, rethinking reality with a pencil and a restless mind—you are the reason ripple-based science is possible. You are the next generation of **cosmic listeners**.

The Structured Vacuum Itself

To the silent field that holds every memory, bends with every intention, and never asks to be noticed—this entire framework is simply **your whisper, remembered.**

And finally, to the **observer within** each of us—

You are not here to watch. You are here to remember. Thank you for listening.

About the Author

Sandeep Chavan is an independent physicist, educator, and founder of the *Universal Energy Dynamics (UED)* framework—a radical reinterpretation of reality that replaces particles with ripples, forces with consequences, and matter with memory. With over a hundred published works spanning theoretical physics, philosophy, cosmology, and human behavior, Sandeep has emerged as a unique voice rewriting modern science from first principles.

An engineer by training and a teacher by calling, Sandeep has spent decades simplifying complex scientific ideas for students in classrooms across India. His signature approach merges rigorous logic with intuitive clarity, earning him a reputation as both a challenger of orthodoxy and a cultivator of future minds.

Sandeep's work is not institutional—it is **existential**. He builds theories not for acceptance, but for alignment. From questioning Einstein's spacetime to rethinking the nature of atoms, his research is grounded in the belief that **the universe is not built—it is remembered.** The ripple, not the particle, is the key.

This book marks the beginning of a new direction in chemistry—one that dissolves electrons, redefines bonding, and places the atom in the context of cosmic harmony. For Sandeep, atoms are not objects. They are **echoes.**

He lives, writes, and thinks from a quiet corner of the world—yet his ripples now resonate across it.

To know an atom is not to dissect it. To know it is to feel its memory in motion.

Don't miss out!

Visit the website below and you can sign up to receive emails whenever Sandeep Chavan publishes a new book. There's no charge and no obligation.

https://books2read.com/r/B-A-EPGPC-TWWPG

BOOKS 2 READ

Connecting independent readers to independent writers.

Did you love *Chavan's Atomic Model*? Then you should read *$E \neq mc^2$: The Collapse of Classical Assumptions*[1] by Sandeep Chavan!

[2]

What if everything we thought we knew about energy is incomplete?

For over a century, $\mathbf{E = mc^2}$ has symbolized the elegance of physics—a compact equation that married mass with energy and reshaped the modern scientific worldview. But what if it also marked the beginning of a deeper misunderstanding?

In *$E \neq mc^2$: The Collapse of Classical Assumptions*, independent physicist and philosopher **Sandeep Chavan** presents a provocative new vision of reality—one where energy isn't stored in mass, light isn't emitted from particles, and explosions are not releases, but realignments.

---

1. https://books2read.com/u/4DaMLe

2. https://books2read.com/u/4DaMLe

At the heart of this transformation is **Chavan's Realignment Equation**, a field-centric model that replaces classical concepts of mass, force, and emission with ripple-based coherence, structured vacuum dynamics, and a new limit on perception: the **Alignment Threshold Time (ATT)**. What Einstein treated as constants—mass, light speed, time—are reinterpreted as consequences of ripple synchronization within an intelligent field.

This book isn't just theoretical—it rewrites the practical language of energy, light, and decay. From **fire and nuclear bombs** to **the Big Bang** and **black holes**, each phenomenon is re-explored through the lens of **Universal Energy Dynamics (UED)**. Along the way, it challenges the foundations of particle physics, space-time models, and object-based reality.

Designed for scientists, engineers, philosophers, and curious thinkers alike, $E \neq mc^2$ blends rigorous insight with radical clarity. It features:

A full breakdown of the **Realignment Equation**Field-based alternatives to mass, inertia, and forceThe **Collapse Law**: why things vanish, not explodeImplications for measurement, engineering, and consciousnessA bold rethinking of **Einstein's legacy**—not discarded, but upgraded

This is not science fiction. It is a **post-Einstein doctrine** built on consequence, perception, and the hidden geometry of energy.

Whether you're a physicist looking for a new frontier, a student questioning the standard model, or a philosopher seeking a bridge between matter and meaning, this book opens a door into a universe that doesn't move—it **realigns**.

Nothing glows. Everything echoes. Nothing moves. Everything adjusts.

It's time to collapse old assumptions—and perceive energy as it truly is.

Read more at https://www.sandeepjchavan.com/.

Also by Sandeep Chavan

Chavan's UED research
CMGED Research Report
Chavan's Cosmic Continuum (C^3)
Frequency
Ripples of Reality
Reality Realigned
The Great Realignment
I – The Cosmic Leverage
Chavan's Cosmic Theory of Relativity (CTR)
Chavan's Theory of Light
Feeling Like a Physicist
TIME – The New Definition
$E \neq mc^2$: The Collapse of Classical Assumptions
The Chavan Codex
Chavan's Atomic Model
Chavan's Theory of Everything

Echoes of War: The Russia-Ukraine Conflict and Its Global Ripples
Shattered Lives
The War Economy
Borders Redrawn

Arms and Allies
Collateral Damage
Justice in the Ashes

Somewhere in the Darkness of the Cosmos
The Blueprint of the Universe
The Cosmic Breath
From the Edge of the Universe

The Eternal Sage Chronicles
The Sage and the Sovereign
The Family Pillar
The People's Flame
The Warrior's Soul
The Bridging Mentor
The Guiding Light

The Exodus
The Last Cycle: 2075

The Fundamental Theory of Life & Existence
Energy in Action: The Soul Demystified
The Active Mind
Energy in Connection
Death: The Disconnection
Subconsciousness: The Chosen Disconnection

Desires: The Engine of Life

The Restoration of Human Knowledge

Restoring Bhagavad Gita

The War of Thoughts

Cracks in the Known

The Human Game

The Divine Dilemma

The Original Intent

The Answers

Ultra Modern Life Philosophy

The Duality of Laughter and Tears

Take It Easy

Compare to Accept

Standalone

It's Not AI, It's AHI - Amplified Human Intelligence

The Decision Paradox

The Four Sapiens

Malicious Script of Indian Polity

The IIT Legacy & Global Impact

The IIT Legacy & Global Impact

You are just a Version of Your Original

Win the Game You Didn't Choose

Always Keep Your Bags Packed
Reveal to Shield - Learn the Game of Facades
Engage Beyond Elections
Culture, Identity & Change: The Evolution of Indian Society
The Triangular Dynamics
Never Trust Without Doubt
If You Can't Change the Country, Leave the Country
Live Fearless with Fear
India: The Investment Magnet
Realignment
Powering a Billion Dreams
Universal Skill
Extension: From Brush to Algorithm
Human Side of Einstein
The Future Knows No Lines
Let's Bring the Generations to the Table
Unified Energy Dynamics
A Beat That Never Dies
Strategic Horizons
Right From Scratch
The Forgotten Innovators
Physics Fixing
The Savior Illusion
The Selfish Side
The Age of the Weakest Sapiens
Neural Wars
The Absorption: There Is No Death
Strange Therapist
Bubbles: Become a Drop
Beyond the Blue
Half-Scholar
The Law of Truth
Failure

Philosophy 2.0: A Structural Manifesto
Chaos: A Gentle Guide to the Unresolvable
Self-Help: New Syllabus
Rippling
Game: How to Play the Share Market
Out of Syllabus
Saffronized: The Code Behind the Crown
The Anger
Moksha: The Liberation
Chavan vs Einstein: The Universe Was Never Curved
Rocket Science: The Ripple Gambling
The Cosmos: The Still Universe
The Taste of Silence
Chavan's Post-Materialist Human Model
10-Second Rule
Chavan vs Hawking
It was Never a Wave Signal
Quantum Courtroom: The Trial of a Century
Chavan–Heisenberg Ripple Uncertainty Principle
The Great Disruption
Dark ≠ Hidden
Breath
Calculus of Life
The Earth Doesn't Move
Black Hole Don't Exist
Chavan vs Kant
Naked Truth: Everything and Else
The Moon: Co-Author of Life's Script
World War-3

Watch for more at https://www.sandeepjchavan.com/.

www.ingramcontent.com/pod-product-compliance
Lightning Source LLC
LaVergne TN
LVHW090318210926
848480LV00001B/284
9798231633746